Prespacetime Journal | December 2020 | Volume 11 | Issue 7

Prespacetime Journal

Volume 11 Issue 7
December 2020

Universal Bifurcations, Measurement Problem, Cosmological Model, & Special Tubular Surface

Editor:
Huping Hu, Ph.D., J.D.

Editors-at-Large:
Philip E. Gibbs, Ph.D.
Dainis Zeps, Ph.D.

Advisory Board

ISSN: 2153-8301   Prespacetime Journal   www.prespacetime.com
Published by QuantumDream, Inc.

(Publishd in Prespacetime Journal | December 2020 | Volume 11| Issue 7 | pp. 582-641)
Table of Contents

i

Table of Contents

Articles

GR Articles

Math-Phys Article

Opinion

Commentary

ISSN: 2153-8301   Prespacetime Journal   www.prespacetime.com
Published by QuantumDream, Inc.

Article

CMB Anomalies from Self-Organized Criticality

Ervin Goldfain[*]

Advanced Technology and Sensor Group, Welch Allyn Inc., Skaneateles Falls, NY 13153

Abstract

The power spectrum of the cosmic microwave background (CMB) quantifies the distribution of relic radiation left over from the early Universe. As of today, CMB data acquired by Planck and WMAP satellites exhibit certain anomalies that challenge the standard model of cosmology (ΛCDM). The goal of this brief report is to sketch up an intriguing connection between CMB anomalies and self-organized criticality (SOC). Our proposal bypasses the interpretation of CMB anomalies based on Loop Quantum Cosmology (LQC).

Keywords: CMB anomalies, power spectrum, ΛCDM, self-organized criticality, loop quantum cosmology.

1. Introduction

ΛCDM predicts that the primordial power spectrum is nearly scale-invariant and described by the power law [1]

$$P_R(k) = A_s \left(\frac{k}{k_*}\right)^{n_s - 1} \tag{1}$$

in which k is the wavenumber measured in Mpc^{-1}, A_s denotes the amplitude of the scalar mode of spectral index n_s and k_* represents the so-called pivot mode. According to the LQC model, (1) acquires a *suppression factor* $f(k)$ whose effect is negligible for large wavenumbers, that is $k >> k_0$,

$$f(k) = O(1), \ k >> k_0 \tag{2}$$

[*]Correspondence: Ervin Goldfain, Ph.D., Photonics CoE, Welch Allyn Inc., Skaneateles Falls, NY 13153, USA
E-mail: ervingoldfain@gmail.com

with k_0 being a reference value. By contrast, suppression of the nearly scale-invariant spectrum (1) occurs if the wavenumber drops down near $k \leq 10k_0$. The modified primordial spectrum predicted by LQC is given by [1]

$$P_R(k) = f(k) A_s \left(\frac{k}{k_*}\right)^{n_s-1} \tag{3}$$

There is more than one way to display and simulate the power spectrum (1). For example, [2] brings up the following set of scaling relations inspired by LQC and inflation

$$P_R(k) = \begin{cases} A_s \left(k/k_*\right)^{n_s-1}, \; k > k_* \\ A_s \left(k/k_*\right)^{q}, \; k_I < k \leq k_* \\ A_s \left(k_I/k_*\right)^{q} \left(k/k_I\right)^{2}, \; k \leq k_I \end{cases} \tag{3}$$

where k_* and k_I are the characteristic scales at primordial cosmological times and q is an exponent dissimilar in magnitude to the spectral index n_s.

The goal of this short report is to explore a scenario where (2) and (3) derive from a radically different approach to the CMB formation. Demanding *self-similarity* in the complex dynamics of large structures implies that CMB is the outcome of a *global SOC process*. Our preliminary analysis is consistent with earlier proposals where SOC is conjectured to assume a critical role in astrophysics and cosmology [3-7].

2. Mathematics of SOC: a short overview

Consider a large-scale ensemble of observables undergoing a second-order phase transition. The transition is driven by the control parameter λ as it approaches the critical value λ_c. Near the critical point and for systems of infinite extent ($L \rightarrow \infty$), the correlation length ξ diverges as

$$\xi \sim (\lambda - \lambda_c)^{-\nu} \; ; \; L \rightarrow \infty, \; \lambda \rightarrow \lambda_c \tag{4}$$

In the transition region, a relevant variable of the system is also a diverging quantity which scales as

$$A_\infty(\lambda) \sim |\lambda - \lambda_c|^{-\varsigma} \; ; \; L \rightarrow \infty, \; \lambda \rightarrow \lambda_c \tag{5}$$

where ζ is a critical exponent. In what follows, we introduce the notation

$$\tau_s = -(\zeta/\nu) \tag{6}$$

There are two distinct cases associated with the power-law (4). If the size of the system greatly exceeds the correlation length, $L \gg \xi$, by (4) and (5) we write

$$A_L(\lambda) \sim \xi^{-\tau_s} \; ; \; (L \gg \xi, \lambda \to \lambda_c) \tag{6}$$

In the opposite case, $L \ll \xi$, the system size takes over the scaling behavior and (6) turns into

$$A_L(\lambda) \sim L^{-\tau_s} \; ; \; (L \ll \xi, \lambda \to \lambda_c) \tag{7}$$

Taken together, (6) and (7) define the *finite-size scaling* (FSS) ansatz

$$A_L(\lambda) = \xi^{-\tau_s} \Phi(L/\xi) \; ; \; (L \to \infty, \lambda \to \lambda_c) \tag{8}$$

where the scaling function controls the finite-size effects of critical behavior and is defined as

$$\Phi(x) = \begin{cases} const; \; |x| \gg 1 \\ x^{-\tau_s} \; ; \; x \to 0 \end{cases} \tag{9}$$

To transition from the framework of critical phenomena to SOC, one simply identifies the correlation length with the concept of *avalanche-size*, i.e.,

$$s = \xi \; ; \quad s_c = L \tag{10}$$

The probability distribution defining the FSS ansatz in SOC is a natural extrapolation of (8) and takes the form of a probability distribution [8]

$$P(s,L) \sim s^{-\tau_s} \Phi(s/s_c) \text{ for } s \gg 1, \; L \gg 1 \tag{11a}$$

$$s_c(L) \sim L^{D_0} \text{ for } L \gg 1 \tag{11b}$$

in which τ_s and D_0 are called the *avalanche-size exponent* and the *avalanche dimension*, respectively. Quite generally, (11) shows that, for a system of finite extent and large size avalanches, the avalanche-size probability behaves as a fractal function times a generic scaling function. To enable all moments of (11) to exist, the scaling function must decay sufficiently fast. One obtains the following representation of the scaling function upon power expanding it around zero,

$$\Phi(x) \sim \begin{cases} \Phi(0) + \Phi'(0)x + \dfrac{1}{2}\Phi''(0)x^2 + ..., & x \ll 1 \\ \to 0, & x \gg 1 \end{cases} \tag{12}$$

The avalanche-size probability must be normalized to unity and its average be diverging along with $L \to \infty$, which leads to the following constraints

$$\sum_{s=1}^{\infty} P(s;L) = 1 \qquad \text{for } L < \infty, \tag{13}$$

$$\langle s \rangle = \sum_{s=1}^{\infty} sP(s;L) \to \infty \ \text{ for } L \to \infty \tag{14}$$

Under the assumption that $\Phi(0) \neq 0$, the behavior of (11) for an infinite system size may be approximated as

$$\lim_{L \to \infty} P(s;L) \sim s^{-\tau_s} \Phi(0) \tag{15}$$

Furthermore, to comply with (13) and (14), the avalanche-size exponent must fall in the range

$$1 < \tau_s \leq 2 \tag{16}$$

It is important to note that, while SOC has a clear *statistical* underpinning as described by (11), the power spectra (1)-(3) are based on *deterministic* measurements unrelated to probabilistic assumptions. A helpful analogy between (1)-(3) and (11) is nevertheless possible, with the caveat that (16) is not necessarily relevant insofar as the CMB spectrum is concerned.

With these considerations in mind, we set up next a parallel between (1)-(3) and the slowly driven evolution of SOC towards a *non-equilibrium steady state*.

3. CMB as non-equilibrium steady state

Since it is always convenient to work with dimensionless entities, we normalize the wavenumbers entering (1)-(3) according to

$$k^0 = \frac{k}{K}, \ \ k_*^0 = \frac{k_*}{K} \tag{17}$$

in which K stands for a suitably chosen reference value. Comparative inspection of (1)-(3) and (11) suggests the term-by-term identification

$$s^0 = \frac{k^0}{k_*^0}, \quad k^0 = s^0 k_*^0 = \frac{s^0}{s_c^0} \tag{18}$$

$$s_c^0 = (k_*^0)^{-1} = \frac{K}{k_*} \tag{19}$$

Following [8], the critical avalanche size (19) scales with the maximal extension of the wavenumber space (or ultraviolet cutoff) Δ_{UV} as in

$$s_c^0 \sim (\Delta_{UV})^{D_0} \tag{20}$$

Labeling the scaling function and avalanche-size exponent by, respectively,

$$\Phi(\frac{s^0}{s_c^0}) = A_s\, f(k^0), \quad \tau_s = 1 - n_s \tag{21}$$

enables one to cast (2)-(3) in the same form as (11), namely,

$$\boxed{P_R(s^0) = s_0^{-\tau_s}\, \Phi(\frac{s^0}{s_c^0}), \quad \text{for } \Delta_{UV} \gg 1,\; s^0 \gg 1} \tag{22a}$$

$$\boxed{s_c^0 \sim (\Delta_{UV})^{D_0}, \quad \text{for } \Delta_{UV} \gg 1} \tag{22b}$$

Note that the scaling function $\Phi(...)$ is nearly constant for $s^0 \ll s_c^0$ in the limit of unbounded wavenumbers $\Delta_{UV} \to \infty$ [8]. By default, this condition corresponds to the regime of infinitesimal spatial separations in the CMB map, where the spectra (1) and (2) are nearly scale-invariant.

We close with few remarks that (in our view) are important for future extensions of this work:

1) among the most straightforward scaling functions $\Phi(...)$ that may be considered in simulations are the Heaviside step-function and the exponential function, where the latter characterizes so-called "branching SOC processes" [8].
2) the CMB angular power spectrum displayed in [9] falls off at large multipole moments in a strikingly similar manner with the data collapse of [8, page 284]. Does this analogy further supports our approach or does it arise from a different rationale altogether?

3) To gain credibility, follow-up extensions of these ideas must successfully recover the large-scale power anomaly described by the parameter $S_{1/2}$, as well as the lensing amplitude A_L derived in [1].

Received August 12, 2020; Accepted December 21, 2020

References

1. https://arxiv.org/pdf/2001.11689.pdf
2. https://arxiv.org/pdf/2005.01796.pdf
3. http://article.scirea.org/pdf/16004.pdf
4. http://article.scirea.org/pdf/16005.pdf
5. https://link.springer.com/article/10.1007/s11214-018-0489-2
6. https://www.researchgate.net/publication/280104885_GRB_Cosmology_and_Self-organized_Criticality_in_GRBs
7. https://www.researchgate.net/publication/343386149_Solving_the_Flatness_and_Horizon_Problems_via_Self-Organized_Criticality
8. Christensen K. and Moloney N. R., "Complexity and Criticality", Imperial College Press 2005.
9. http://folk.uio.no/hke/AST5220/v11/AST5220_2_2011.pdf

Article

Three Generations & Fermion Chirality from Universal Bifurcations

Ervin Goldfain[*]

Advanced Technology and Sensor Group, Welch Allyn Inc., Skaneateles Falls, NY 13153

Abstract

The Standard Model (SM) fails to account for either the triplication of fermion families or chiral symmetry breaking in the electroweak sector. Here we show that both phenomena arise from the approach to chaos of quantum theory near the Fermi scale.

Keywords: Standard Model, fermion generations, chiral symmetry breaking, Feigenbaum route to chaos, dissipative maps, period-doubling bifurcations.

1. Introduction

Iterated maps of the unit interval are generic models of dynamical systems in discrete time [7-8]. The standard representation of these models is based on first order difference equations having the form

$$x_{n+1} = f(x_n, \lambda) \tag{1}$$

where $n = 1, 2, ...$ is the iteration index and λ a control parameter. The dynamics expressed by (1) can be either *conservative* or *dissipative*. In the former case, the function (1) is monotonic and describes a one-to-one mapping, whereas in the latter case is non-monotonic and describes a two-to-one mapping. Typical examples of dissipative systems include the *quadratic map* and *unimodal maps* [1-5]. In 1978, Feigenbaum has discovered that the onset of chaos in quadratic maps occurs through *period-doubling bifurcations* driven by changes of the control parameter [1-2]. It was later shown that the period-doubling transition to chaos with the same universal attributes develops in many multi-dimensional dissipative nonlinear systems [3-5] In particular, unimodal maps of the form [8-9]

$$f_\lambda(x) = \lambda f(x) \tag{2}$$

<hr>

[*]Correspondence: Ervin Goldfain, Ph.D., Photonics CoE, Welch Allyn Inc., Skaneateles Falls, NY 13153, USA
E-mail: ervingoldfain@gmail.com

$$x \in [-1,1] \; ; \; f(0) = 1 \; ; \; \lambda < 1 \tag{3}$$

exhibit the following behavior: for small values of λ, (2) has a single stable fixed point and all nearby points converge to it under multiple iterations given by (1). Ramping up λ to a critical value λ_1 makes the fixed point unstable and produces a new stable pair of points of period 2. Further increasing λ to another value (λ_2) bifurcates this cycle into a cycle of period 4. The bifurcation process continues with a sequence of cycles of period 2^j , $j \geq 3$, eventually leading to a Cantor set structure that attracts almost all the points of the interval $[-1,1]$. On letting λ increase beyond an endpoint value λ_N, $N \gg 1$, stable periodic orbits surface again and split up in a similar way. In the new sequence, λ scans another series of critical values corresponding to cycles of period $3 \cdot 2^j$, $j = 0,1,2,...$ and so on [8-9].

2. Chaotic behavior of quantum theory near the Fermi scale

Quantum Mechanics contains many instances where unimodal functions of the type (2) show up. For example, a quantum wave packet initially centered at $x = 0$ [10]

$$\psi(x,0) = \exp\left(-\frac{x^2}{2}\right) \tag{3}$$

evolves in time according to

$$\psi(x,t) = \exp\left(-\frac{i\omega t}{2}\right)\psi(x,0) = \exp\left(-\frac{i\omega t}{2}\right)\exp\left(-\frac{x^2}{2}\right) \tag{4}$$

Another example is a wavefunction localized in x space as in [10]

$$\psi(x) = \frac{\sin K(x - x_0)}{\pi(x - x_0)} \; , \; K \gg 1 \tag{5}$$

whose momentum space representation is

$$\varphi(k) = \frac{\exp(-ikx_0)}{\sqrt{2\pi}} \tag{6}$$

Gauge theory demands the description of physical phenomena related to (4) or (6) to be independent of any arbitrary phase factor $\exp(-i\chi)$ added to either one of them. With reference to (4), a global gauge transformation defined by the complex parameter $\lambda_c = \exp(-i\chi)$ assumes the form

$$\psi'(x,t) = \lambda_c \psi(x,t) = \exp(-i\chi)\psi(x,t) \tag{7}$$

Following a standard procedure, we pass from Lorentzian to Euclidean coordinates using the prescription

$$\chi = \omega t' = \omega(-it'_E) \tag{8}$$

$$\chi_E = -\omega t'_E \tag{9}$$

The Euclidean control parameter in the small angle approximation may be presented as

$$\lambda_E = \exp(-\omega t'_E) \approx 1 - \omega t'_E, \quad |\omega t'_E| = \varepsilon << 1 \tag{10}$$

In the context of the *minimal fractal manifold* (MFM) [11], (10) highlights the connection between the Euclidean control parameter λ_E and the scale-dependent deviation from spacetime dimensionality $\varepsilon(\mu) = 4 - D(\mu) << 1$. One concludes from (10) that the behavior of (7) is controlled by a parameter that runs with the energy scale as in $\lambda_E = \lambda_E(\mu)$.

Taken together, all these observations indicate that (7) undergoes period-doubling transition to chaos upon letting λ_E scan a continuous range of values. In particular, as shown in [6],

1) Gauge bosons develop from the 2^j bifurcation pattern, while fermions from the $3 \cdot 2^j$ pattern. *The first stage of the fermion sector* $(j = 0)$ *contains a triplication of fermion generations*, in full agreement with the experimental basis of the SM.

2) The neutrino-antineutrino branch developed at $j = 0$ displays an intrinsic chiral asymmetry upon applying the CPT operator, which entails that only one of the Left (L) and Right (R) neutrino states exist. *It follows that chiral symmetry breaking is rooted in the neutrino sector and stems from the transition to chaos driven by the global gauge transformation (7).*

Both findings are consistent with [11-12], where the SM is conjectured to represent a *self-contained multifractal set,* whose composition is constrained by the so-called "sum-of-squares" relationship. The violation of chiral symmetry falls in line with spacetime anisotropy induced by fractional differential and integral operators near or above the Fermi scale [13].

Received August 17, 2020; Accepted December 21, 2020

References

1. M. J. Feigenbaum, "Quantitative universality for a class of nonlinear transformations", J. Stat. Phys. 19:25–52 (1978).
2. M. J. Feigenbaum, "The universal metric properties of nonlinear transformations", J. Stat. Phys. 21:669–706 (1979).
3. P. Cvitanovic (ed.), "Universality in Chaos", 2nd Ed. (Adam Bilger, Boston, 1989).
4. P. Collet, J.-P. Eckmann, and H. Koch, "Period Doubling Bifurcations for Families of Maps on Rn", J. Stat. Phys. 25:1–14 (1981)
5. http://spkuz.narod.ru/2005JSP.pdf
6. E. Goldfain, "A Bifurcation Model of the Quantum Field", Physica A 165: 399-419 (1990). A copy of this reference is available at:
   https://www.academia.edu/38767540/A_bifurcation_model_of_the_quantum_field
7. P.Collet, J.-P. Eckman, "Iterated Maps on the Interval as Dynamical Systems", (Birkhäuser, Boston, 1980)
8. E. A. Jackson, "Perspectives of nonlinear dynamics", (Cambridge Univ. Press, 1991).
9. K. J. Falconer, "The geometry of fractal sets", (Cambridge Univ. Press, 1988).
10. D. Bohm, "Quantum Theory", (Dover, New York, 1989).
11. http://www.aracneeditrice.it/aracneweb/index.php/pubblicazione.html?item=9788854889972
   https://www.researchgate.net/publication/278849474_Introduction_to_Fractional_Field_Theory_consolidated_version
12. https://www.researchgate.net/publication/343426122_Derivation_of_the_Sum-of-Squares_Relationship
13. https://www.sciencedirect.com/science/article/abs/pii/S1007570406001183

Article

Proposal for Solving the Quantum Measurement Problem

Amal Pushp [1] & **Shad Azmi**[2]
[1] Dept. of Physics, School of Basic Sci., Manipal Univ., Rajasthan, India
[2] Dept. of Math., Aligarh Muslim University, Uttar Pradesh, India

Abstract

The measurement problem is one of the greatest unsolved mystery of the foundations of quantum mechanics. This paper is an exploration of the underlying subtleties of this phenomenon from the standpoint of the quantum mechanical interpretations that are available to us at the moment. First, we revisit the infamous double slit experiment and contrast between the observer dependent and observer independent effects on the measurement process. In this regard, we have tried to establish a causal relationship between the observer's consciousness and the collapse of the wave function. The point is justified through the application of Godel's incompleteness theorems. Finally, we perform a thought experiment which shall validate our conclusions.

Keywords: Wave function collapse, superposition, observer effect.

1 Introduction and Summary

The measurement problem dates back to the initial days of quantum theory. Attempts to solve it include models of wave function collapse, several interpretations of quantum mechanics, theory of decoherence etc [1-14]. However, no single interpretation has been able to completely explain the various irregularities of the quantum world to complete satisfaction. Unfortunately, there is a lack of consensus for a single formalism, there exists between certain groups of physicists, a favouritism for a particular interpretation which sadly, is loaded with their own prejudices and biases. In science, calculations and interpretation should be kept on equal footing. Extreme notions such as shut up and calculate" must be avoided. A certain balance must coalesce the two into a complete and true picture of reality. In this paper, we keep our investigations grounded along these lines with the prime motive to carve out a clearer picture of the measurement problem. We have tried, to keep a check on our biases and have allowed the empirical evidence(s) lead the way, from which we have tried to map our ideas onto these empirical evidences.

2 Double Slit Experiment Revisited

Things happen pretty classically with the double slit experiment for mundane objects which seem to obey our Boolean logic and evolutionary intuitions, however things get more complicated as one transitions to the quantum realm. The Boolean logic of black or white breakdowns to give way to a much more balanced approach to thinking. Here we describe the experiment itself along with the actual experimental observations as empirical evidence. A stream of electrons passing through a double slit creates an interference pattern at the screen, i.e. they exhibit wave like properties as evident from the de Broglie hypothesis. Even if electrons are shot one at a time, it still results into an interference pattern which is startling [15, 16]. If we imagine what is really happening here, we would see, that the electron leaves as a particle, mutates into a probability wave, goes through both slits, interferes with itself and finally hits the screen.

---

[1] Corresponding author: E-mail: amal.201006004@muj.manipal.edu

This seems outrageous! Probing into the underlying mathematical underpinnings would reveal that the scenario here is that the electron which is a wave packet with waves of different frequencies superposed and propagating with a group velocity, goes through both slits as different parts of the same probability wave and then recombines to form a localized particle and hits the screen. The equations further reveals that all the different possibilities of the electron going through the double slit is in superposition. If the electron is viewed as a probability wave, we have,

$$\psi(x) = \alpha\psi_a(x) + \beta\psi_b(x) \tag{2.1}$$

which represents the superposition state and the probability density is given by,

$$P(x) = \alpha\psi_a(x) + \beta\psi_b(x)^2 = \alpha\psi_a(x)^2 + \beta\psi_b(x)^2 + \alpha\beta^*\psi_a(x)\psi_b^*(x) + \alpha^*\beta\psi_a^*(x)\psi_b(x) \tag{2.2}$$

The above equation actually represents the quantum interference. Now, if we put a detector inorder to really know which slit the electron went through, the observations are breathtaking. We model the scenario by drawing contrast between the observer dependent (detector present) and observer independent (detector still present) effects. A thought experiment is also performed to visualize and clarify the conclusions.

CASE I (observer + detector): As mentioned above, quantum particles such as electrons or for that matter, things such as Bucky balls when shot towards a double slit, an interference pattern is observed which happens in the absence of a measuring device and which is simply impossible if the particles behave classically since the particle must have gone through a single slit and a single detector at an instant. Only waves exhibit the interference phenomenon and not classical particles such as cars, stars or any other macroscopic objects. In the case of macroscopic objects, the interference is not significant enough to be measured or detected. Classical particles are bounded in their own potential wells which constrains their behaviour to that of a localised particle whereas this is not true for objects which belong to the set of the quantum realm which unlike the classical particles exhibit a non local phenomenon therefore, behave as that of a wave which in turn explains the interference phenomenon. This clearly differentiates the two domain of reality, and establishes the impossibility of classical particle showing interference pattern. This leads to the conclusion that the electrons must have behaved like that of a wave while passing through the two slits. Naturally, this leads to the verification of this claim. In the case of quantum particles, if one of the detector is removed, the consequences remain unchanged and still there would be no interference phenomenon.

CASE II (only detector): Consider the situation where the observer is not present or is present indirectly in a way that he is not watching the experiment through his eyes. In this condition, as described by the mathematics, the electron passes through both the slits and detectors but still the interference pattern is absent. How come? This is strange! When the observer now observes the system, he sees the electron striking only one of the two detectors which indicates its definite location. In this case the detector is basically replacing the observer. At this point a question may arise as to why then the interference pattern disappears in the absence of an observer? The answer lies in the fact that the detector is already acting as a hindrance in the path of the electron causing the particle to change its behaviour.

According to Einstein-Podolsky-Rosen (EPR), physical reality is dependent on the ability to measure a system without disturbing it in any way [17]. In view of the aforementioned discussions on the observer dependent and observer independent effects, the situation is quite the opposite. The experimental observations are indicating without room for slightest doubt, that the conscious observer is playing a fundamental role in the measurement process and it is in regard with this fact that we now argue that there's a causality relation between the observer's consciousness and the collapse of the particle from a superposition of possibilities to an ontic state. This is achieved through the application of Godel's incompleteness theorems [18]. The real mystery in resolving the measurement problem is that we are lacking an ontology, a mechanism through which the transition from abstract mathematical construct to a physically realistic state is actually taking place. The transition (or collapse) can be shown explicitly as follows. We start with a superposition state of the form, say,

$$|\Psi\rangle = \sum_\Phi |\Phi\rangle \, \Phi\Psi \tag{2.3}$$

The above equation uses the identity operator $\hat{I} = \sum_\Phi |\Phi\rangle \, \Phi$ in some fixed orthonormal basis $\left\{ |1\rangle, |2\rangle, ..., |\Phi\rangle \right\}$ as follows:

$$|\Psi\rangle = \hat{I}|\Psi\rangle = \sum_\Phi |\Phi\rangle \, \Phi\Psi = \sum_\Phi (\Phi\Psi)|\Phi\rangle = \sum_\Phi a_\Phi |\Phi\rangle \tag{2.4}$$

where the quantum probability amplitudes have been set as, $a_\Phi = \Phi\Psi$. The Born rule can be used to compute the probability density, given by,

$$P(\Phi) = \Psi\hat{\mathbb{P}}|\Psi\rangle = \Psi\Phi\Phi\Psi = a_\Phi^* a_\Phi = a_\Phi{}^2 \tag{2.5}$$

An act of observation (measurement) would reduce the superposition state to a definite eigenstate state represented by, $\sum_\Phi a_\Phi |\Phi\rangle \rightarrow |\Phi\rangle$ and now the state of the system at any future time would be determined by the Schrodinger evolution,

$$i\hbar\frac{d}{dt}(|\Phi\rangle) = \hat{H}(|\Phi\rangle) \tag{2.6}$$

where $\hat{H}$ is the Hermitian operator called the Hamiltonian. More details can be found in [19].

Godel's first incompleteness theorem states that [20], If T is a computably axiomatized, consistent extension of N, then T is undecidable and hence incomplete". In other words [21], Any consistent formal system F within which a certain amount of elementary arithmetic can be carried out is incomplete; i.e., there are statements of the language of F which can neither be proved nor disproved in F. The second theorem states that [21], For any consistent system F within which a certain amount of elementary arithmetic can be carried out, the consistency of F cannot be proved in F itself".

In [22], it was shown that Godel's theorems are applicable in the domain of physics in general and quantum theory in particular. Building on their results, we incorporate Godel's theorems into our system and establish a causal relationship between consciousness and the collapse of the wave function. To contrast this, let us show a proof by contradiction in which we put forward the proposition that, there exists no causal relationship between consciousness and the collapse of the quantum wave function. CASE II tells us that in the absence of a conscious observer, the particle passes through both the slits and is recorded by both the detectors as inferred from the mathematical calculations. However, when a conscious being is introduced into that bounded system, the wave function of the electron collapses and it strikes only one of the detectors. In other words, the particle is in a quantum superposition and is not localised until observed. This is a contradiction with our proposition, hence there must exist a causal relationship between consciousness and the collapse of the wave function. Godel's first theorem explains the main point here. If one cannot prove the influence of a conscious observer on the quantum measurement process, this does not in any case necessitates the falsehood of the claim. Here we take the causal relationship as an a priori statement in the form of an axiom which is well supported by the theorems proposed by Kurt Godel and then deducing the results which map onto the experimental evidences which have been performed and recorded over the past century since the dawn of this quantum enterprise itself and in some cases, long before that as well.

It is also found that Cochran's argument that elementary particles like electrons possess a rudimentary level of consciousness, a kind of self activity, supports our ideas and thoughts presented in the paper [23]. We now discuss what decoherence has to say about the measurement problem and why it doesn't really solve the problem [24]. We clarify the actual role played by decoherence in the process.

3   Decoherence and the Measurement problem

Proponents of the decoherence theory argue that it explains the collapse of the particle. The explanation goes along the following lines. The particle which is to be measured is interacted with another particle which disturbs the state of the first particle as part of the interaction process. As a result, the collapse of the first particle happens due to the particle which has been used to interact with it. But if this is the case, an attentive reader would ask then what is the reason behind the collapse of the measuring particle? We would need another measuring device to collapse it and then another apparatus to collapse the previous measuring device and this goes on and on in a chain called the von Neumann chain. The von Neumann chain can be explained mathematically in a beautiful way using the following array of equations [25],

$$|O\rangle\,(|M_o\rangle\,|F_o\rangle\,|W_o\rangle) = (|A\rangle + |B\rangle)\,(|M_o\rangle\,|F_o\rangle\,|W_o\rangle) \xrightarrow{t_1} \tag{3.1}$$

$$(|A\rangle\,|M_A\rangle + |B\rangle\,|M_B\rangle)\,(|F_o\rangle\,|W_o\rangle) \xrightarrow{t_2} \tag{3.2}$$

$$(|A\rangle\,|M_A\rangle\,|F_A\rangle + |B\rangle\,|M_B\rangle\,|F_B\rangle)\,(|W_o\rangle) \xrightarrow{t_3} \tag{3.3}$$

$$(|A\rangle\,|M_A\rangle\,|F_A\rangle\,|W_A\rangle + |B\rangle\,|M_B\rangle\,|F_B\rangle\,|W_B\rangle) \tag{3.4}$$

The above equation has taken into account the evolution of a system $|O\rangle = |A\rangle + |B\rangle$ and its measurement by Wigner and his friend. $|M_o\rangle$ is the initial state of the measuring device associated with the system. Now, an act of conscious observation by Wigner would collapse the superposition in the eqn. (6) to either of the states and thus the measurement would yield either $|A\rangle\,|M_A\rangle\,|F_A\rangle$ or $|B\rangle\,|M_B\rangle\,|F_B\rangle$.

On careful analysis, it is realised that the decoherence theory is basically talking about the mechanism of the collapse that would supposedly take place naturally by environment induced decoherence. Decoherence would collapse a particle naturally if left out for some period of time, maybe years. On the other hand, von Neumann's interpretation talks about an instantaneous collapse brought about by conscious observation. Where the former is talking about the how", the latter is answering the why", by establishing the necessary causality between the two phenomena which we showed earlier. Fortunately/Unfortunately, questions of such sort are philosophically laden but that does not imply that they are irrelevant to the progress of the scientific enterprise since they might act as an overarching guiding principle. Sometimes, one has to contend with such philosophically laden ideas and questions when grappling with them because they are, after all intimately entangled" together.

4   A Thought Experiment

Here we perform a simple thought experiment which is a variation of a similar one performed in the case of Schrodinger's cat [26]. We model the double slit experiment scenario in terms of a circuit diagram as shown in the figure below. In our model, the switch replaces the poison and bulb replaces the cat. The switch in this case acts naturally when electron wave passes through both the slits simultaneously. Then, using quantum logic gates, we analyse the information flow between the switch and the bulb in order to verify its correlation with the theoretical arguments and whether or not they are consistent within the framework. We start from the classical case. If classical balls are used instead of quantum particles, there would be no interference pattern. The circuit actually makes no sense in this particular case. We are mainly interested in how this classical behaviour transforms to the quantum regime.

Coming to the quantum regime, the bulb would glow if the circuit is fully complete. The state of switch-bulb system can be represented by the following entangled state,

$$|\Psi\rangle = \frac{1}{\sqrt{2}}\,(|0\rangle \otimes |on\rangle + |1\rangle \otimes |off\rangle) \tag{4.1}$$

Also unless a rational quantum CNOT gate is performed between the switch and the bulb, the final state of the system would be given by the statistical mixture [26],

$$\rho = a^2\,|0\rangle\,0 \otimes |on\rangle\,on + b^2\,|1\rangle\,1 \otimes |off\rangle\,off \tag{4.2}$$

ISSN: 2153-8301   Prespacetime Journal   www.prespacetime.com
Published by QuantumDream, Inc.

In the absence of a measuring device, an electron would pass from both the slits simultaneously, the circuit would be completed without any problem, the bulb would glow and thus an interference pattern would be formed as a consequence. Coming to the observer dependent case, the electron passes either from the first or from the second slit and not both simultaneously. As a result, the circuit remains incomplete, the bulb wouldn't glow and also the interference pattern vanishes. Note that we haven't made any major change in the system and have just introduced the observer into the picture. If we assume a priori that the consciousness model is correct as an axiom, the results we observe emerge automatically and they are in full agreement with empirical evidences as well as our thought experiment as presented here.

Lastly, we are left with the observer-independent case and which is also perhaps the most intriguing part. In this case, although the electron would pass through both the slits simultaneously, the circuit would be completed and the bulb would also glow but still the interference pattern would vanish. It is important to recall that this is something which is only perceived from the math and it is only when an actual observation is made by a conscious observer that the electron seems to strike a single detector which would also imply that it passes through a single slit rather than both simultaneously. This would mean that the circuit is not completed and also the bulb wouldn't glow when a conscious observation is made. A straightforward contradiction can be seen here. There is severe lack of correlation between what the math says and what is actually seen upon observation. A denial of the role of conscious observation at this point, if any, would seem to be brought about by force.

5   Conclusion

While researching for our paper, we got accustomed to some of the work other people have done or are doing along the similar lines, such as Penrose [27], Stapp [28] and Irwin [29] alongside many others. This paper is our introductory paper addressing the measurement problem. We have tried to make this paper self contained and elaborate at the same time just like other quantum physical objects- which is somewhat relevant afterall. We are looking forward to carry research along these lines in the future and also invite others who are thinking like us to join this journey.

(Published in Prespacetime Journal | December 2020 | Volume 11 | Issue 7 | pp. 592-600)

Pushp, A., & Azmi, S.,*Proposal for Solving the Quantum Measurement Problem*

Figure 1: Circuit Diagram: The figure shows an electron source through which electrons would initiate and pass through the double slits. Behind the slits, a circuit has been constructed consisting of a bulb. If electrons pass through both slits simultaneously, the bulb would glow otherwise it would remain off.

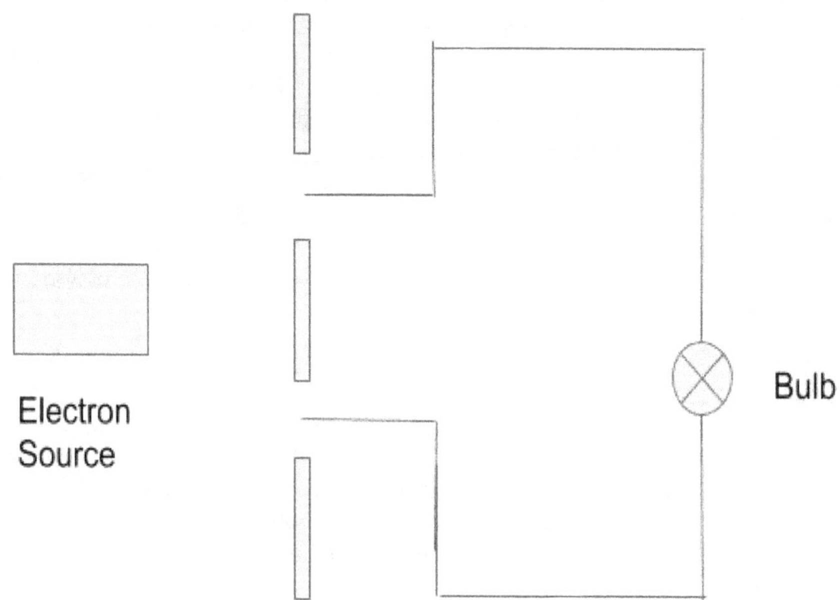

Figure 2: Quantum Network Model: The model reveals the information flow in the thought experiment. Part 1 of the figure represents the entangled switch-bulb system. Bulb in the initial state $|on\rangle$ is linked to the switch via a quantum CNOT gate resulting in the state given by eqn. (11). In the second part, the switch evolves freely to $a\,|0\rangle + b\,|1\rangle$ and is coupled to the bulb via a classical information channel.

6 Acknowledgement

We would like to thank Vikram Zaveri and Andrew Knight for insightful discussions, their valuable suggestions alongside critical judgements.

Received November 17, 2020; Accepted December 21, 2020

References

[1] A. Bassi, K. Lochan, S. Satin, T.P. Singh and H. Ulbricht, *Models of Wave-function Collapse, Underlying Theories and Experimental Tests*, Rev. Mod. Phys. **85 (471)**, (2013). arXiv:https://arxiv.org/abs/1204.4325

[2] N. Bohr, *The Quantum Postulate and the Recent Development of Atomic Theory*, Nature. **121** 580590, (1928).

[3] W. Heisenberg, *ber den anschaulichen Inhalt der quantentheoretischen Kinematik und Mechanik*, Z.Phys. **43** 172-198, (1927). Translated by John Wheeler and Wojciech Zurek, in *Quantum Theory and Measurement*, (1983).

[4] David Bohm, *A Suggested Interpretation of the Quantum Theory in Terms of "Hidden Variables", I*, Physical Review. (1952), **85**, pp 166179.

[5] David Bohm, *A Suggested Interpretation of the Quantum Theory in Terms of "Hidden Variables", II*, Physical Review. (1952), **85**, pp 180193.

[6] Hugh Everett, *Relative State Formulation of Quantum Mechanics*, Rev. Mod. Phys. **29(3)** 454-462, (1957).

[7] David Wallace, *The Emergent Multiverse: Quantum Theory According to the Everett Interpretation.* Oxford University Press (2012).

[8] C. Rovelli, *Relational Quantum Mechanics*, International Journal of Theoretical Physics **35**; 1996: 1637-1678. `arXiv:quant-ph/9609002`.

[9] Ghirardi, G.C., Rimini, A., and Weber, T. *Unified dynamics for microscopic and macroscopic systems*, Phys. Rev. D. **34(2)**: 470491, (1986). `doi:10.1103/PhysRevD.34.470`

[10] Roger Penrose, *On Gravity's role in Quantum State Reduction*, Gen. Rel. Gravit. **28(5)**: 581600, (1996). `doi:10.1007/BF02105068`

[11] Christopher A Fuchs, Rdiger Schack. *A Quantum-Bayesian Route to Quantum-State Space* Found. Phys. **41(3)**: 345356, (2010). `doi:10.1007/s10701-009-9404-8` `https://arxiv.org/abs/0912.4252`

[12] John von Neumann, *Mathematical Foundations of Quantum Mechanics*, Princeton University Press, (1932).

[13] Eugene Wigner, Henry Margenau, *"Remarks on the Mind Body Question, in Symmetries and Reflections, Scientific Essays"*, Am. J. Phys. **35(12)**, 1169-1170, (1967).

[14] Maximilian Schlosshauer. *"Decoherence, the measurement problem, and interpretations of quantum mechanics"*, Rev. Mod. Phys. **76(4)**, 12671305, (2005). `arXiv:quant-ph/0312059`

[15] P. G. Merli, G. F. Missiroli and G. Pozzi. *On the statistical aspect of electron interference phenomena*, Am. J. Phys. **44** , 306-307, (1976).

[16] A. Tonomura, J. Endo, T. Matsuda, T. Kawasaki and H. Ezawa. *Demonstration of single-electron build-up of an interference pattern*, Am. J. Phys. **57**, 117-120, (1989).

[17] A. Einstein, B. Podolsky, N. Rosen. *Can Quantum-Mechanical Description of Physical Reality Be Considered Complete?*, Phys. Rev. **47(10)**, 777-780, (1935).

[18] K. Godel. *ber formal unentscheidbare Stze der Principia Mathematica und verwandter Systeme, I*, Monatshefte fr Mathematik und Physik. **38(1)**, pp 173-198, (1931).

[19] A. Pushp, *The Dead-Alive Physicist Gedankenexperiment Debunked*, to appear.

[20] Y. Khomskii, *Godels Incompleteness Theorem*, https://www.math.uni-hamburg.de/home/khomskii/recursion/Goedel.pdf

[21] P. Raatikainen, *Godel's Incompleteness Theorems* https://plato.stanford.edu/entries/goedel-incompleteness/

[22] John M. Myers, F. Hadi Madjid. *Incompleteness theorem for physics*, `arXiv:1803.10589`

[23] A. A. Cochran, *Relationships between quantum physics and biology* , Found. Phys. **1(3)**, 235-250, (1971).

[24] E. Joos, *Elements of Environmental Decoherence*, `arXiv:quant-ph/9908008`

[25] A. Knight, *Macroscopic Quantum Superpositions Cannot Be Measured, Even in Principle*, `philarchive`, (2020).

ISSN: 2153-8301   Prespacetime Journal   www.prespacetime.com
Published by QuantumDream, Inc.

[26] R. Ionicioiu, *Schrodinger's Cat: Where Does The Entanglement Come From?*, Quanta **6**, 57-60, (2017). `arXiv:1603.07986`

[27] R. Penrose, *The Road to Reality: A Complete Guide to the Laws of the Universe*, (2004).

[28] H. Stapp, *Mind, Matter and Quantum Mechanics* , (The Frontiers Collection) Springer (2009).

[29] K. Irwin, *A New Approach To The Hard Problem of Consciousness*, Activitas Nervosa Superior **62(25)**, (2020).

ISSN: 2153-8301   Prespacetime Journal   www.prespacetime.com
Published by QuantumDream, Inc.

Article

Cosmic Acceleration of Modified Holographic Ricci Dark Energy Cosmological Model in Teleparallel Gravity

V. M. Raut[*]

Dept. of Mathematics, Shri Shivaji Science College, Amravati, India

Abstract

In this paper, the author considers FRW metric filled with matter and modified holographic Ricci dark energy with volumetric power law towards the gravitational field equations for the depiction model. The geometrical and physical parameters of the model are studied.

Keywords: FRW metric, $f(T)$ gravity, cosmic acceleration, Ricci dark energy, teleparallel gravity.

1. Introduction

The universe is experiencing a phase of accelerated expansion is confirmed by multiple observations [1-2]. The source of this acceleration is usually attributed to an exotic type of fluid with negative pressure called commonly dark energy.Currently, there are two main approaches explaining this accelerated expansion. One way is to construct different dark energy candidates and the other way is modification of Einstein's theory of gravitation.The modified theories of gravity, i.e., $f(R), f(G), f(R,T), f(T)$ and scalar tensor theory have gained a lot of interest to explain the nature of DE [3-4].

One framework that has gained momentum in recent years is that of teleparallel gravity (TEGR) [5-6] and its modification, the $f(T)$ teleparallel theory [7-9].In $f(T)$ gravity, the Teleparallel Lagrangian density described by the function of torsion scalar T in order to account for the late time cosmic acceleration [10-13] .

The potential works dealing with thermodynamics, stability and ΛCDM model in $f(T)$, are that of Bamba [14] and Setare [15]. Indeed, Bamba and collaborators widely explored thermodynamics of the apparent horizon in $f(T)$ gravity with equilibrium and non-equilibrium descriptions [14], obtaining interesting results.Several Relativists [16-35] explored the expansion of the universe with different techniques in $f(T)$ gravity.

---

[*]Correspondence: V. M. Raut, Department of Mathematics, Shri Shivaji Science College, Amravati-444603, India
E-mail: vinayraut18@gmail.com

Motivated by the above investigations, in this paper, FRW universe filled with interacting Dark matter and Modified Holographic Ricci Dark Energy (MHRDE) has been studied within the framework of $f(T)$ theory.

2. A brief review of $f(T)$ cosmology & MHRDE

The action is defined by generalizing the teleparallel theory, i.e., $f(T)$ theory as

$$S = \int \left[T + f(T) + L_{matter} \right] e \, d^4 x .$$

(1)

Here $f(T)$ denotes an algebraic function of the torsion scalar T. Making the functional variation of the action of equation (1) with respect to tetrads , we get the following equations of motion

$$S_\mu^{\nu\rho} \partial_\rho T f_{TT} + \left[e^{-1} e_\mu^i \partial_\rho (e e_i^\alpha S_\alpha^{\nu\rho}) + T_{\lambda\mu}^\alpha S_\alpha^{\nu\lambda} \right](1 + f_T) + \frac{1}{4} \delta_\mu^\nu (T + f) = T_\mu^\nu ,$$

(2)

where T_μ^ν is the energy momentum tensor , $f_T = \dfrac{df(T)}{dT}$. The energy momentum tensors for matter and MHRDE are defined as

$$T_{\mu\nu} = \rho_m u_\mu u_\nu \; ; \overline{T}_{\mu\nu} = (\rho_\lambda + p_\lambda) u_\mu u_\nu - g_{\mu\nu} p_\lambda ,$$

(3)

where ρ_m is the energy densities of matter, ρ_λ is the holographic dark energy and p_λ is the pressure of the holographic dark energy. The field equations (2) for the energy momentum tensor for matter and holographic dark energy becomes

$$S_\mu^{\nu\rho} \partial_\rho T f_{TT} + \left[e^{-1} e_\mu^i \partial_\rho (e e_i^\alpha S_\alpha^{\nu\rho}) + T_{\lambda\mu}^\alpha S_\alpha^{\nu\lambda} \right](1 + f_T) + \frac{1}{4} \delta_\mu^\nu (T + f) = T_\mu^\nu + \overline{T}_\mu^\nu .$$

(4)

3. Metric and the field equations

The FRW line element represented by the following metric

$$ds^2 = dt^2 - a^2(t) \left\{ \frac{dr^2}{1 - kr^2} + r^2(d\theta^2 + \sin^2\theta d\phi^2) \right\} ,$$

(5)

where the $0 \le \theta \le \pi$ and $0 \le \phi \le \pi$ are the azimuthal and polar angles of the spherical co-ordinate system . k represent the curvature of the space. In this work $k = 0$ is considered i.e. the flat universe. The energy–momentum tensor of dark energy can be parameterized as

$$\overline{T}_{\nu\mu} = diag\left[1, -w_\lambda, -w_\lambda, -w_\lambda\right]\rho_\lambda,\tag{6}$$

where $w_\lambda = \dfrac{p_\lambda}{\rho_\lambda}$ is the equation of state (EoS) parameter of dark energy and ρ_m, ρ_λ are the energy densities of matter and dark energy and p_λ is the pressure of the dark energy. The expression of the energy density of dark energy is given by

$$\rho_\lambda = \left(3\eta_1 \ddot{H}H^{-1} + 3\eta_2 \dot{H} + 3\eta_3 H^2\right),\tag{7}$$

where η_1, η_2, η_3 are the arbitrary dimensionless parameters. For the line element (5), using equations (3), (6) and (7), the field equation (4) for matter and MHRDE in FRW metric takes the following form

$$\frac{\dot{a}}{a}\dot{T}f_{TT} + \left[\frac{\ddot{a}}{a} + 2\frac{\dot{a}^2}{a^2}\right](1 + f_T) + \frac{1}{4}(T + f) = -\omega_\lambda\rho_\lambda,\tag{8}$$

$$3(1 + f_T)\frac{\dot{a}^2}{a^2} + \frac{1}{4}(T + f) = (\rho_m + \rho_\lambda),\tag{9}$$

where the overhead dot represents differentiation with respect to time.

4. Solutions of the field equations

Equations (8) and (9) are two highly non-linear differential equations with six unknowns namely $a, f, T, \omega_\lambda, \rho_\lambda, \rho_m$. The system is thus initially undetermined. Thus there is a need of extra physical conditions to solve the field equations completely. Firstly, the scale factor of the form is chosen $a = t^n$. Secondly, the simplest and commonly considered depiction $f(T)$ model is considered as $f(T) = T^\psi$. The Hubble parameter is obtained as $H = \dfrac{\dot{a}}{a} = \dfrac{n}{t}$. The Scalar expansion is given by $\theta = 3H = \dfrac{3n}{t}$. The expansion scalar and the Hubble parameter H vanish with the increase of time [36-44]. The deceleration parameter is

$$q = \frac{d}{dt}\left(\frac{1}{H}\right) - 1 = \frac{1}{n} - 1.\tag{10}$$

The Universe is accelerating for $n > 1$ which leads to the fact that the deceleration parameter 'q' is negative. The MHRDE density is given by

$$\rho_\lambda = \frac{(6\eta_1 - 3n\eta_2 + 3n^2\eta_3)}{t^2}. \tag{11}$$

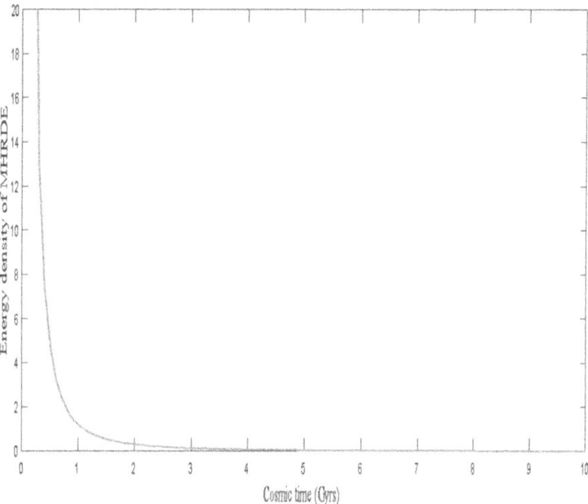

Figure 1.Energy density of MHRDE vs time.

The energy density of matter is given by

$$\rho_m = \frac{(3n^2 - 6\eta_1 + 3n\eta_2 - 3n^2\eta_3)}{t^2}. \tag{12}$$

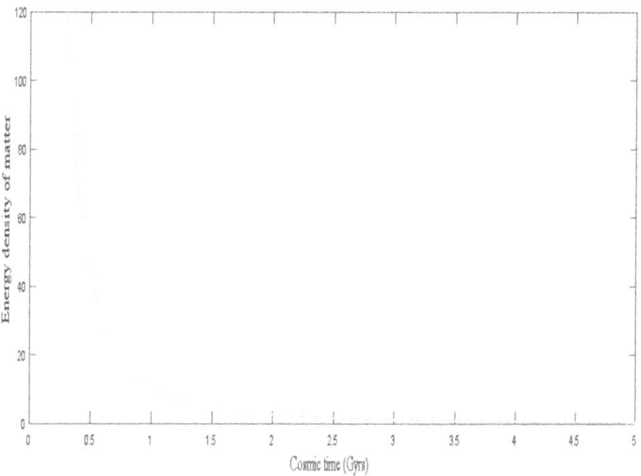

Figure 2. Energy density of matter vstime .

ISSN: 2153-8301   Prespacetime Journal   www.prespacetime.com
                              Published by QuantumDream, Inc.

The equations of state parameter of MHRDE is given by

$$\omega_\lambda = -\frac{(3n^2 - 2n)}{(6\eta_1 - 3n\eta_2 + 3n^2\eta_3)}.$$

(13)

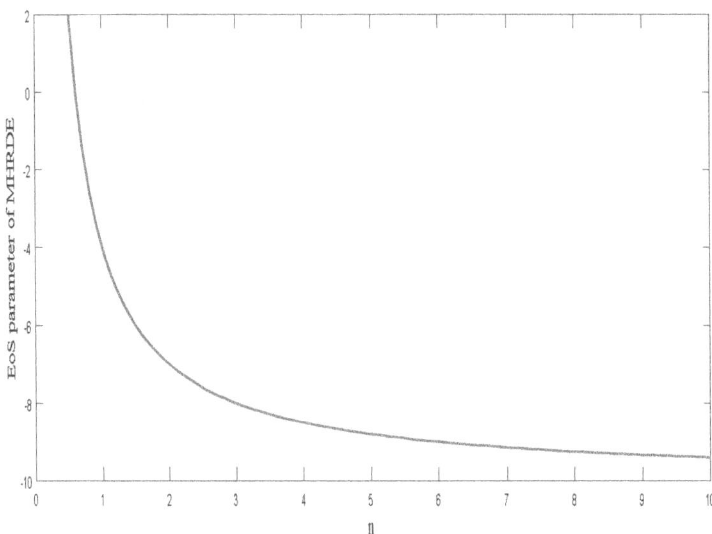

Figure 3. EoS Parameter vs*n*.

The energy densities of MHRDE and matter are decreasing functions of cosmic time [45-55].Itis observed that both ρ_λ and ρ_m decreases throughout the evolution of model and vanishes at late time's leadsthe volume expansion of the universe as depicted in figure 1 and 2.In the derived model as shown in figure 3, at an early stage the EoS parameter is positive i.e. at an initial epoch the universe is dominated by matter era. After a very short interval, the universe enters in DE phase.

The matter density parameter Ω_m and HDE density parameter Ω_λ are given respectively by

$$\Omega_m = \frac{\rho_m}{3H^2} = \frac{(3n^2 - 6\eta_1 + 3n\eta_2 - 3n^2\eta_3)}{3n^2}.$$

(14)

$$\Omega_\lambda = \frac{\rho_\lambda}{3H^2} = \frac{(6\eta_1 - 3n\eta_2 + 3n^2\eta_3)}{3n^2}.$$

(15)

The overall density parameter as

$$\Omega = \Omega_m + \Omega_\lambda = 1.$$

(16)

It is observed that the overall density parameter has a constant value 1.Therefore, the derived MHRDE model predicts that the anisotropy will dampout and the universe will become isotropic.

5. Conclusions

In this paper, FRW metric with the energy momentum tensor for matter and MHRDE with volumetric power law within the framework of $f(T)$ gravity has been considered. It is observed that the shear scalar is decreasing function of time and ultimately tends to zero as $t \to \infty$. MHRDE energy density and matter energy density decrease as the cosmic time increases.

The deceleration parameter indicates that the universe is accelerating which is consistent with the present day observations that universe is undergoing the accelerated.The transitional behavior of the EoS parameter of this model fits with SNIa data [56-58].The total density parameter is equal to 1 achieves isotropy which is reproducible with current observations. The derived solution is stable against perturbation of graviton field.

Received October 2, 2020; Accepted November 15, 2020

References

1.   A. G. Riess et al. : Astron. J. 116, 1009 (1998)
2.   S. Perlmutter et al.:Astrophys. J. 517, 565 (1999)
3.   M. Kowalski et al.:Astrophys. J. 686, 749 (2008) .
4.   S.D.H. Hsu: Phys. Lett. B 594 , 13(2004).
5.   A. Einstein 1928, Sitz. Preuss. Akad. Wiss. p. 217; ibid p. 224
6.   M. Tsamparlis, Phys. Lett. A 75, 27 (1979)
7.   G.R. Bengochea and R. Ferraro, Phys. Rev. D 79, 124019 (2009)
8.   E.V. Linder, Phys. Rev. D 81, 127301 (2010)
9.   M. Kr˘s˘sa´k and E.N. Saridakis, Class. Quantum Grav. 33, 115009 (2016)
10.   Bengochea, G.R and Ferraro, R.: Phys. Rev. D79,124019(2009).
11.   Bamba, K. ,Geng, C.Q. and Lee, C.C.: arXiv: [astro-ph.] 1008.4036 (2010).
12.   K. Bamba and C. Geng J. Cosmol. Astropart. Phys. 11 ,008, (2011) .
13.   R. Myrzakulov Eur. Phys. J. C. 71 ,1752, (2011).
14.   K. Bamba and C. Q. Geng, JCAP 1111, 008 (2011) [arXiv:1109.1694 [gr-qc]].
15.   M.R. Setare, N. Mohammadipour, JCAP 01 (2013) 015 [arXiv: 1301.4891].
16.   Linder, E.V.: Phys. Rev. D81,127301(2010).
17.   K. Karami, A. Abdolmaleki: arXiv:1009.3587v1 [physics.gen-ph] 18 Sep 2010.
18.   B.Li, T.P. Sotiriou and J.D. Barrow: arXiv:1103.2786v1 [astro-ph.CO] 14 Mar 2011.
19.   R.J.Yang: arXiv:1007.3571v3 [gr-qc] 9 Nov 2011.
20.   M. Sharif and S. Rani: arXiv:1105.6228v1 [gr-qc] 31 May 2011.
21.   Jamil, M., Momeni, D., Myrzakulov, R.:Eur. Phys. J. C. 72, 2122 (2012)
22.   Jamil, M., Yesmakhanova, K., Momeni, D., Myrzakulov, R.: Cen. Eur. J. Phys.10,1065(2012)
23.   Setare, M., Darabi, F.:GenRelativGravit44, 2521 (2012)
24.   Setare, M., Houndjo, M.:Can. J. Phys. 91, 168 (2013)
25.   C. Geng, C. Lee, E. Saridakis, Y. Wu Phys. Lett. B 704 384(2011)

26. C. Geng, C. Lee, E. Saridakis JCAP 1201 ,002(2012)
27. M.E.Rodrigues,A.V.Kpadonou,F.Rahaman,P.J.OliveiraandM.J.S. Houndjo : arXiv:1408.2689v1 [gr-qc] 12 Aug 2014
28. M. Sharif and S.Azeem: Commun. Theor. Phys. 61 , 482–490(2014).
29. M. Jamil and M. Yussouf : arXiv:1502.00777v1 [gr-qc] 3 Feb 2015.
30. G. Abbas, A.Kanwal, M. Zubair: Astrophys Space Sci (2015) 357:109.
31. M. Zubair,S.Waheed: Astrophys Space Sci (2015) 360:68.
32. A.Jawad,S. Chattopadhyay,,S.Rani: Astrophys Space Sci (2016) 361:231.
33. S.Capozziello,R.D'Agostino,O.Luongo: Gen RelativGravit (2017) 49:141.
34. S.Capozziello, O.Luongo, R.Pincak,A.Ravanpak: Gen RelativGravit (2018) 50:53.
35. V.J. Dagwal and D. D. Pawar : Modern Physics Letters A Vol. 34 (2019) 1950357 (19 pages).
36. A.Y.Shaikh,A.S.Shaikh,K.S.Wankhade:J. Astrophys. Astr. (2019) 40:25.
37. Katore S. D., Adhav K. S., Shaikh A. Y., Sancheti M. M.2011, Astrophys. Space Sci., 333, 333
38. Katore S. D., Adhav K. S., Shaikh A. Y., Sarkate N. K. 2010,Int. J. Theor. Phys., 49, 2558
39. S.D.Katore, A.Y.ShaikhPrespacetime J., 3(11),1087(2012).
40. S.D.Katore, A.Y.Shaikh. Bul. J. Phys., 39, 241(2012)
41. S.D.Katore, A.Y.Shaikh. :Afr. Rev. Phys., 9, 0035(2014)
42. S.D.Katore, A.Y.Shaikh: Rom. J. Phys., 59(7–8),715(2014)
43. S.D.Katore, A.Y.Shaikh.:Astrophys. Space Sci.,357(1), 27(2015)
44. S.D.Katore, A.Y.Shaikh, Bulg. J. Phys., 41, 274(2014)
45. S.D.Katore, A.Y.Shaikh., Bhaskar S. 2014, Bulgarian J.Phys., 41(1)
46. Katore S. D., Shaikh A. Y., Kapse D. V., Bhaskar S. A. 2011,Int. J. Theor. Phys., 50, 2644
47. S.D.Katore, A.Y.Shaikh: Int. J. Theor. Phys., 51,1881(2012).
48. S.D.Katore, A.Y.Shaikh:Afr. Rev. Phys., 7, 0004(2012).
49. S.D.Katore ,A.Y.Shaikh:Afr. Rev. Phys., 7, 0054(2012).
50. A.Y.Shaikh: Adv. Astrophys., 2(3)(2017)
51. A.Y.Shaikh,S.D.Katore:Pramana J. Phys., 87, 83(2016)
52. A.Y.Shaikh,S.D.Katore:Pramana J. Phys., 87, 88(2016)
53. A.Y.Shaikh, K.S.Wankhade: Phys. Astron. Int. J.,1(4), 00020(2017)
54. A.Y.Shaikh ,K.S.Wankhade :Theoretical Phys., 2(1)(2017)
55. A.Y.Shaikh,S.V.Gore,S.D.Katore: New Astronomy 80, 101420(2020).
56. A.Y.Shaikh and K.S.Wankhade : arxiv.1912.08044(2019)
57. A.Y. Shaikh, A.S. Shaikh, K.S. Wankhade: arxiv:2006.12300(2020).
58. A.Y.Shaikh and B.Mishra : Int. Jou. ofGeom.Methods in Modern Physics (2020).Doi.org/10.1142/S0219887820501583.

ISSN: 2153-8301   Prespacetime Journal   www.prespacetime.com
Published by QuantumDream, Inc.

Article

Bulk Viscous Cosmological Model in $f(T)$ Gravity by Hybrid Expansion Law

V. M. Raut[*]

Dept. of Mathematics, Shri Shivaji Science College, Amravati, India

Abstract

In this paper, the author studies plane symmetric cosmological model filled with bulk viscosity in the framework of Teleparallel Gravity by using Hybrid Expansion Law. Using depiction model of $f(T)$ gravity, the behavior of accelerating universe is discussed. The physical and kinematical properties of the derived model have been discussed in detail.

Keywords: Plane symmetric, $f(T)$ gravity, bulk viscous, hybrid expansion, cosmological model.

1. Introduction

Recent astrophysical data [1-4] strongly indicate that the universe is accelerating at present. A new kind of matter with positive energy density and with negative pressure dubbed as dark energy is one of the main candidate behind the acceleration of the present day universe. Dark energy occupies about 73% of the energy of our Universe; while dark matter about 23% and the usual baryonic matter 4%. On larger cosmological scales the modification in Einstein-Hilbert action may be an accurate description of a late time cosmic acceleration of the expanding universe. Among the various modifications of Einstein's theory, another one way to look at the theory beyond General Theory of Relativity (GTR) is the Teleparallel Gravity (TG) which is different from GTR (i.e. uses the Weitzenbock connection in place of the LeviCivita connection) which has no curvature but has torsion, responsible for the acceleration of the universe. Ferraro & Fiorini [5] provided models based on modified TG to inflation. In $f(T)$ gravity, the Teleparallel Lagrangian density described by the function of torsion scalar T in order to account for the late time cosmic acceleration [6-9] . Jamil et al. [10] resolved the Dark Matter (DM) problem in the light of $f(T)$ gravity. Setare and Houndjo [11] investigated particle creation in flat Friedman Robertson Walker universe in the framework of $f(T)$ gravity. $f(T)$ gravity has been extensively studied in the literature by several eminent researchers [12-19].

---

[*]Correspondence: V. M. Raut, Department of Mathematics, Shri Shivaji Science College, Amravati-444603, India
E-mail: vinayraut18@gmail.com

Bulk viscosity is the simplest way to study entropy in cosmology. It arises any time a fluid expands rapidly and ceases to be in thermodynamic equilibrium. Eckart [20] developed the first relativistic theory of non-equilibrium thermodynamics to study the effect of bulk viscosity. Krori and Mukherjee [21] explored the evolution of Bianchi cosmologies with bulk viscosity and particle creation. Desikan [22] studied the effect of bulk viscosity for FRW models .Several Relativists considered the behavior of Bulk viscosity in different context [23 -36].

Motivated by above investigations, the paper deals with the investigations of bulk viscous for depiction model of $f(T)$ model by using plane symmetric universe with Hybrid expansion law .

2. Basic set up of $f(T)$ Gravity

Defining the action by generalizing the TG i.e. $f(T)$ theory as

$$S = \int \left[T + f(T) + L_{matter} \right] e\, d^4 x \quad .$$

(1)

Here, $f(T)$ denotes an algebraic function of the torsion scalar T. Making the functional variation of the action (1) with respect to the tetrads, we get the following equations of motion

$$S_{\mu}^{\nu\rho} \partial_{\rho} T f_{TT} + \left[e^{-1} e_{\mu}^{i} \partial_{\rho} \left(e e_{i}^{\alpha} S_{\alpha}^{\nu\rho} \right) + T^{\alpha}{}_{\lambda\mu} S_{\alpha}^{\nu\lambda} \right] (1 + f_T) + \frac{1}{4} \delta_{\mu}^{\nu} (T + f) = k^2 T_{\mu}^{\nu} ,$$

(2)

The field equation (10) is written in terms of the tetrad and partial derivatives and appears very different from Einstein's equations. T_{μ}^{ν} is the energy momentum tensor, $f_T = \dfrac{df(T)}{dT}$ and by setting $f(T) = a_0 = \text{constant}$ this is dynamically equivalent to the General Relativity.

3. Metric and its field equations

The line element of Plane Symmetric space-time is given by

$$ds^2 = dt^2 - A^2 \left(dx^2 + dy^2 \right) - B^2 dz^2 ,$$

(3)

where the metric potentials A and B be the functions of time t only.

The corresponding Torsion scalar is given by

$$T = -2 \left(2 \frac{\dot{A}}{A} \frac{\dot{B}}{B} + \frac{\dot{B}^2}{B^2} \right) .$$

(4)

The energy-momentum tensor T_{ij} for a Bulk Viscous fluid distribution is given by

$$T_{ij} = (\rho + \bar{p}) u_i u_j - \bar{p} g_{ij} ,$$

(5)

ISSN: 2153-8301   Prespacetime Journal   www.prespacetime.com
Published by QuantumDream, Inc.

where

$$\bar{p} = p - \eta u^{i}_{;i} = p - 3\eta H \,,\tag{6}$$

is the effective pressure , η is the coefficient of bulk viscosity , p is the isotropic pressure , ρ is the energy density , $3\eta H$ is usually known as bulk viscous pressure , H is Hubble's parameter and u^{i} is fluid four-velocity vector satisfying

$$u_{i}u^{i} = 1 \,.\tag{7}$$

In the co-moving co-ordinate system, we have from the above equations

$$T_{1}^{1} = T_{2}^{2} = T_{3}^{3} = -\bar{p} \ , \ T_{4}^{4} = \rho \,, T_{i}^{j} = 0 \,, i \neq j \,.\tag{8}$$

Using equations (3) and (8), the field equations of the Teleparallel gravity can be written as

$$\left(T + f(T)\right) - 4\left(1 + f(T)\right)\left\{\frac{\dot{A}^{2}}{A^{2}} + 2\frac{\dot{A}}{A}\frac{\dot{B}}{B}\right\} = 2k^{2}\rho,\tag{9}$$

$$\begin{aligned}4\left\{\frac{\ddot{A}}{A} + \frac{\dot{A}^{2}}{A^{2}} + \frac{\dot{A}}{A}\frac{\dot{B}}{B}\right\}\left(1 + f(T)\right)\\
-16\frac{\dot{A}}{A}\left[\frac{\dot{A}}{A}\left(\frac{\ddot{B}}{B} - \frac{\dot{B}^{2}}{B^{2}}\right) + \left(\frac{\ddot{A}}{A} - \frac{\dot{A}^{2}}{A^{2}}\right)\left(\frac{\dot{A}}{A} + \frac{\dot{B}}{B}\right)\right]f_{TT} - (T + f(T)) = -2k^{2}\bar{p}\end{aligned}\tag{10}$$

where the overhead dot($\cdot$) denotes the derivative with respect to time t.

4. Solutions of the field equations

Here there are two highly non-linear differential equations with eight unknowns namely $A, B, f(T), p, \rho, T, \bar{p}, \eta$. The system is thus initially undetermined. Thus there is a need of extra physical conditions to solve the field equations completely.

The spatial volume for Plane Symmetric space-time is given by

$$V = a^{3} = A^{2}B \,.\tag{8}$$

The expansion scalar:

$$\theta = 3H \,.\tag{9}$$

The mean anisotropy parameter:

$$A_m = \frac{1}{3}\sum_{i=1}^{3}\left(\frac{\Delta H_i}{H}\right)^2 . \tag{10}$$

The shear scalar:

$$\sigma^2 = \frac{1}{2}\left(\sum_{i=1}^{3} H_i^2 - 3H^2\right) = \frac{3}{2} A_m H^2 , \tag{11}$$

where $H_i\,(i=1,2,3)$ represent the directional Hubble parameters in the directions of r,θ,ϕ axes respectively. The hybrid expansion law, for the average scale factor, is given by

$$a(t) = A^2 B = a_1 t^\alpha e^{\beta t} , \tag{12}$$

where α,β are non-negative constants and a_1 is the present value of the scale factor. Equation (12) is known as the hybrid expansion law, which is a combination of a power law and an exponential function. It can be seen that $\alpha = 0$ provides power law cosmology while $\beta = 0$ gives exponential law cosmology. To solve the above set of highly nonlinear equations, the relation between the metric coefficients is considered as

$$A = B^m . \tag{13}$$

For a Baratropic fluid the combined effect of the proper pressure and the Baratropic bulk viscous pressure can be expressed as $\bar{p} = p - 3\eta H = \gamma\rho$, where $p = \gamma_0\rho$, $0 \le \gamma_0 \le 1$ and m is arbitrary constant. Using equations (12) and (13), the metric potentials can be written as

$$B = \alpha_1 t^{\frac{3\alpha}{2m+1}} e^{\frac{3\beta t}{2m+1}} , \qquad \text{where } \alpha_1 = a_1^{\frac{3}{2m+1}} \tag{14}$$

$$A = \alpha_2 t^{\frac{3m\alpha}{2m+1}} e^{\frac{3m\beta t}{2m+1}} , \text{ where } \alpha_2 = a_1^{\frac{3m}{2m+1}} . \tag{15}$$

This is a point type singularity since directional scale factors vanish at initial time.

5. Physical and Kinematical properties

The spatial volume is obtained as

$$V = a_1^3 t^{3\alpha} e^{3\beta t} . \tag{16}$$

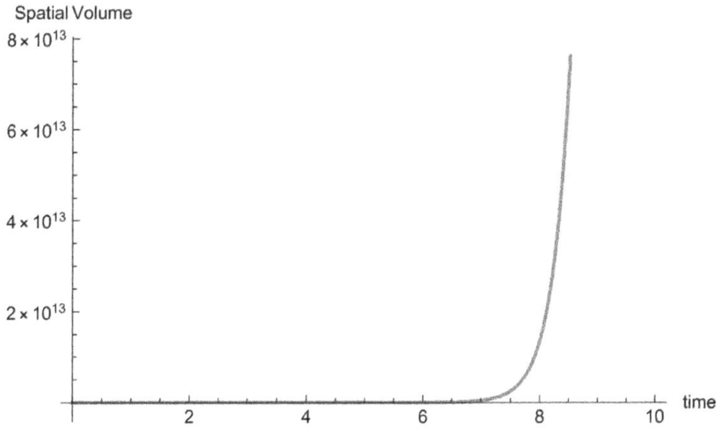

Figure 1: Spatial Volume vs time.

The mean Hubble parameter is found out to be

$$H = \left(\beta + \frac{\alpha}{t} \right). \tag{17}$$

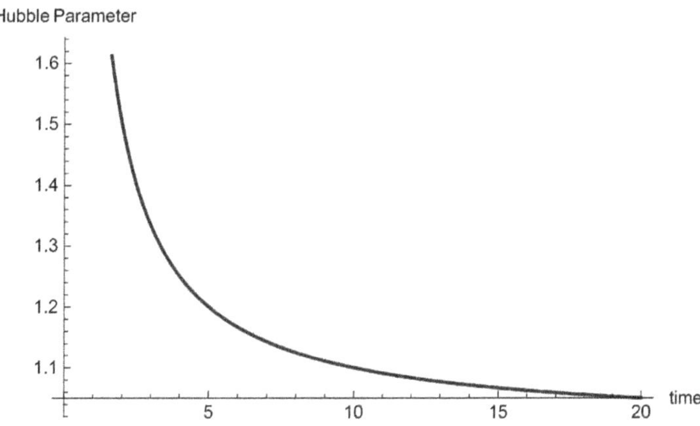

Figure 2: Hubble Parameter vs time.

The Expansion scalar is obtained as

$$\theta = 3\left(\beta + \frac{\alpha}{t} \right). \tag{18}$$

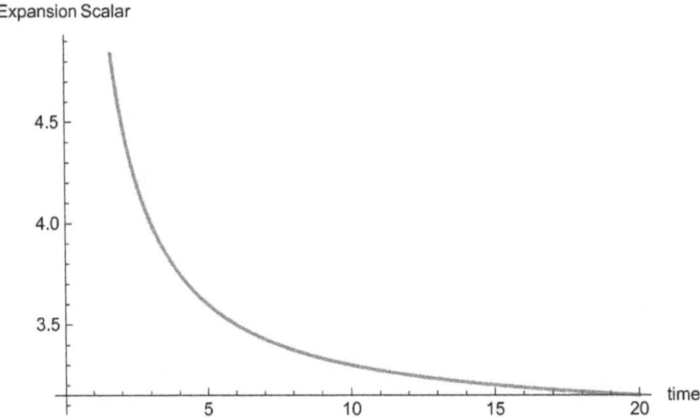

Figure 3. Scalar Expansion vs time.

The mean anisotropic parameter and shear scalar are obtained as

$$A_m = \frac{3(2m^2 + 1)}{(2m + 1)^2}.$$

(19)

$$\sigma^2 = \frac{9(2m^2 + 1)}{2(2m + 1)^2}\left(\beta + \frac{\alpha}{t}\right)^2.$$

(20)

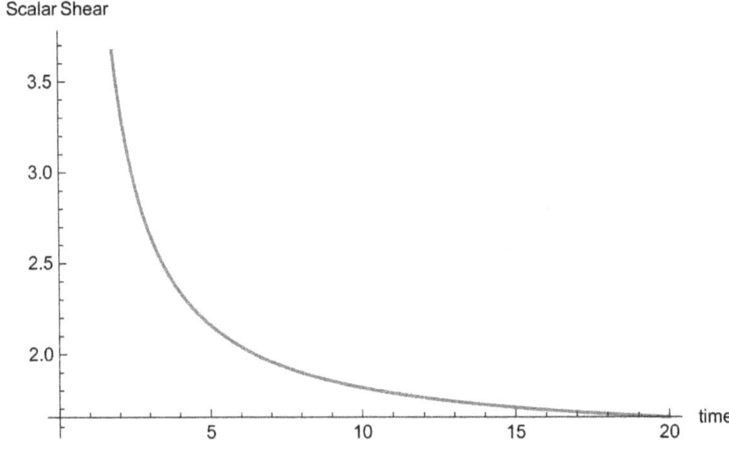

Figure 4. Shear Scalar vs time.

The deceleration parameter is

$$q = \frac{\alpha}{(\beta t + \alpha)^2} - 1.$$

(21)

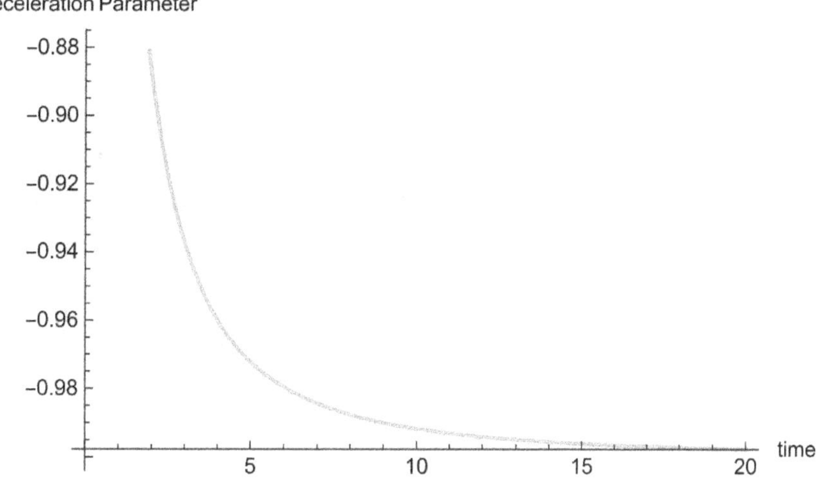

Figure 5. Deceleration Parameter vs time.

Energy density (ρ) of the universe becomes

$$\rho = \left\{ \frac{324\, m^2 (m+2)^2}{(2m+1)^4} \left(\beta + \frac{\alpha}{t} \right)^4 - \frac{36\,(2m+m^2)}{(2m+1)^2} \left(\beta + \frac{\alpha}{t} \right)^2 \right\} \tag{22}$$

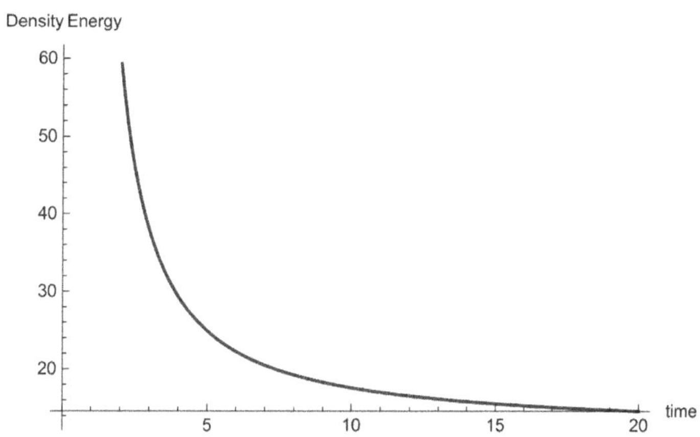

Figure 6. Energy Density vs time.

The isotropic pressure, effective pressure and bulk viscous coefficient respectively are found to be

$$p = \gamma_0 \left\{ \frac{324\, m^2 (m+2)^2}{(2m+1)^4} \left(\beta + \frac{\alpha}{t} \right)^4 - \frac{36\,(2m+m^2)}{(2m+1)^2} \left(\beta + \frac{\alpha}{t} \right)^2 \right\} \tag{23}$$

$$\overline{p} = \gamma \left\{ \frac{324\, m^2 (m+2)^2}{(2m+1)^4} \left(\beta + \frac{\alpha}{t} \right)^4 - \frac{36\,(2m+m^2)}{(2m+1)^2} \left(\beta + \frac{\alpha}{t} \right)^2 \right\} \tag{24}$$

$$\eta = \frac{1}{3H}(\gamma_0 - \gamma)\left\{\frac{324m^2(m+2)^2}{(2m+1)^4}\left(\beta+\frac{\alpha}{t}\right)^4 - \frac{36(2m+m^2)}{(2m+1)^2}\left(\beta+\frac{\alpha}{t}\right)^2\right\}. \qquad (25)$$

Figure 1 depicts that at an initial epoch, the spatial volume vanishes. The Hubble parameter tends to zero i.e. $H \to 0$ as $t \to \infty$ (Figure 2) [37-41]. It is observe that expansion scalar is infinite at $t = 0$ as shown in Figure.3 [42-45]. The shear scalar diverges at an initial epoch as depicted in Figure 4 and tends to zero as $t \to \infty$ [46-49]. The anisotropy parameter tends to a constant, which means that the anisotropy in the universe is maintained throughout. Figure 5 represents the deceleration parameter evolution in time. As $q \approx -1$, the universe is accelerating. It is observed from figure 6 that the energy density decreases as the universe expands [50-58].

6. Conclusions

In this paper, an expansion of the universe in $f(T)$ gravity is explored. An exact solution of the Plane Symmetric space-times using bulk viscous with the help of Hybrid Expansion Law has been investigated. The deceleration parameter appears with negative sign which implies accelerating expansion of the universe. The shear scalar is also infinite at an initial epoch and becomes zero as time increases. It is observed that energy density is positive decreasing function of time and converges to zero as $t \to \infty$.

Received October 3, 2020; Accepted November 15, 2020

References

1. S. Perlmutter et al, Nature (London), 391, 51, (1998)
2. Knop. R et al., Astroph. J., 598, 102 (2003)
3. A.G. Riess et al., Astrophy. J., 607, 665(2004)
4. H. Jassal, J. Bagla and T. Padmanabhan, Phys. Rev. D, 72, 103503 (2005).
5. R. Ferraro & F. Fiorini: Phys. Rev. D, 75, 084031(2007).
6. G.R.Bengochea and R. Ferraro: Phys. Rev. D79, 124019(2009).
7. K.Bamba , C.Q.Geng and C.C.Lee: arXiv: [astro-ph.] 1008.4036 (2010).
8. K. Bamba, C.Q. Geng: JCAP 11, 008 (2011).
9. R. Myrzakulov :Eur. Phys. J. C. 71 ,1752, (2011).
10. M. Jamil, D. Momeni, and R. Myrzakulov, Eur. Phys. J. C 72, 2122 (2012).
11. M. Setare and M. Houndjo, Can. J. Phys. 91, 168 (2013).
12. M.E.Rodrigues,A.V.Kpadonou,F.Rahaman,P.J.Oliveira andM.J.S. Houndjo : arXiv:1408.2689v1 [gr-qc] (2014)
13. M. Jamil and M. Yussouf : arXiv:1502.00777v1 [gr-qc] (2015).
14. G. Abbas, A.Kanwal, M. Zubair: Astrophys Space Sci ,357:109(2015).
15. M. Khurshudyan, R. Myrzakulov and As. Khurshudyan: Modern Physics Letters A Vol. 32, No. 18, 1750097 (2017).

16. M.Hohmann, L.Järv, and U.Ualikhanova: Phys. Rev. D 96, 043508 (2017)

17. S. Capozziello, G. Lambiase,and E. N. Saridakis: Eur Phys J C Part Fields. 77(9): 576(2017).

18. P.Channuie and D.Momeni: arXiv:1712.07927v2 [gr-qc] (2018)

19. A.V. Toporensky, Petr V. Tretyakov: arXiv1911.06064v1 (2019)

20. C Eckart, Phys. Rev. 58, 919 (1940)

21. K D Krori and A Mukherjee, Gen. Relativ. Gra vit. 32, 1429 (2000)

22. K Desikan, Gen. Relativ. Gra vit. 29, 435 (1997)

23. V B Johri and S K Pandey, Int. J. Theor. Phys. 38, 1981 (1999)

24. G P Singh and A Beesham, Aust. J. Phys. 52, 1039 (1999)

25. G P Singh, R V Deshpande and T Singh, Astrophys. Space. Sci. 282, 489 (2002)

26. O Grøn, Astrophys. Space. Sci. 173, 191 (1990)

27. R Maartens, Class. Quantum Grav. 12, 1455 (1995)

28. G P Singh and A Y Kale, Astrophys. Space. Sci. 331, 207 (2011)

29. G P Singh and A Y Kale, Eur. Phys. J. Plus 126, 83 (2011)

30. R Chaubey, Astrophys. Space Sci. 342, 499 (2012)

31. S D Katore, A Y Shaikh, D V Kapse and S A Bhaskar Int. J. Theor. Phys. 50 ,2644(2011).

32. H. Saadat and B. Pourhassan, :Astrophysics and Space Science,vol.344,no.1,pp.237–241,2013.

33. AY Shaikh and S D Katore Bulg. J. Phys. 43,184–194(2016)

34. G.P.Singh and BinayaK.Bishi: Advances in High Energy Physics Volume 2017, Article ID 1390572, 24 pages.

35. S M M Rasouli and P V Moniz, Class. Quantum Grav. 35, 025004 (2018)

36. U K Sharma, R Zia and A Pradhan, J. Astrophys. Astron. 40, Article ID 002 (2019).

37. A.Y.Shaikh,A.S.Shaikh,K.S.Wankhade: J. Astrophys. Astr. (2019) 40:25.

38. Katore S. D., Adhav K. S., Shaikh A. Y., Sancheti M. M.2011, Astrophys. Space Sci., 333, 333

39. Katore S. D., Adhav K. S., Shaikh A. Y., Sarkate N. K. 2010,Int. J. Theor. Phys., 49, 2558

40. S.D.Katore, A.Y.Shaikh Prespacetimc J., 3(11),1087(2012).

41. S.D.Katore, A.Y.Shaikh. Bul. J. Phys., 39, 241(2012)

42. S.D.Katore, A.Y.Shaikh. :Afr. Rev. Phys., 9, 0035(2014)

43. S.D.Katore, A.Y.Shaikh : Rom. J. Phys., 59(7–8),715(2014)

44. S.D.Katore, A.Y.Shaikh.: Astrophys. Space Sci.,357(1), 27(2015)

45. S.D.Katore, A.Y.Shaikh, Bulg. J. Phys., 41, 274(2014)

46. S.D.Katore, A.Y.Shaikh., Bhaskar S. 2014, Bulgarian J.Phys., 41(1)

47. S.D.Katore, A.Y.Shaikh: Int. J. Theor. Phys., 51,1881(2012).

48. S.D.Katore, A.Y.Shaikh :Afr. Rev. Phys., 7, 0004(2012).

49. S.D.Katore , A.Y.Shaikh :Afr. Rev. Phys., 7, 0054(2012).

50. A.Y.Shaikh : Adv. Astrophys., 2(3)(2017)

51. A.Y.Shaikh ,S.D. Katore :Pramana J. Phys., 87, 83(2016)

52. A.Y.Shaikh ,S.D. Katore : Pramana J. Phys., 87, 88(2016)

53. A.Y.Shaikh , K.S.Wankhade : Phys. Astron. Int. J.,1(4), 00020(2017)

54. A.Y.Shaikh ,K.S.Wankhade :Theoretical Phys., 2(1)(2017)

55. A.Y.Shaikh,S.V.Gore,S.D.Katore: New Astronomy 80, 101420(2020).

56. A.Y.Shaikh and K.S.Wankhade : arxiv.1912.08044(2019)

57. A.Y. Shaikh, A.S. Shaikh, K.S. Wankhade: arxiv:2006.12300(2020).

58. A.Y.Shaikh and B.Mishra : Int. Jou. of Geom.Methods in Modern Physics (2020).Doi.org/10.1142/S0219887820501583.

Article

Bianchi Type-VI$_0$ Cosmological Model with Bulk Viscosity & Dust Distribution in C-field Theory

Jaipal Singh[*1], Atul Tyagi[1] & Gajendra Pal Singh[2]

[1]Dept. of Math. & Statistics, Univ. College of Sci., MLS University, Udaipur, India
[2]Dept. of Math., Geetanjali Institute of Technical Studies, Udaipur, India

Abstract

We have investigated Bianchi type-VI$_0$ cosmological model with bulk viscosity and dust distribution in C-field theory. We assume the matter content of the universe is in the form of dust, which leads to p = 0. To get the deterministic model of the universe we assumed $A = B^2$, where A and B are metric potentials. We find that the creation field (C) increase with time which matches with the result of H-N theory. The physical and geometrical aspects of the model are also discussed.

Keywords: Cosmology, C-field, dust, bulk viscosity, Bianchi Type VI$_0$.

1. Introduction

The homogeneous and isotropic models are described by FRW model but in smaller scale, the universe is neither homogeneous and nor isotropic. The study of Bianchi type VI$_0$ cosmological models create more interest as these models are homogeneous and anisotropic. Homogeneous and anisotropic models have been widely studied in the frame work of general relativity by several authors viz. Collins and Hawking [12], Dunn and Tupper [14], Roy and Bali [21], Bali [9], Bali and Singh [8].

Barrow [10] has pointed out the Bianchi type VI$_0$ model of universe and explained cosmological problem. Bali and Goyal[3] investigated Inflationary Scenario in Bianchi type- V Spacetime with Variable Bulk Viscosity and Dark Energy in Radiation Dominated Phase. Ellis and McCollum [15] investigated solution of Einstein field equation from Bianchi type VI$_0$ space time in stiff fluid. Ruban [22] and Collins [13] studied some exact solution of Bianchi type VI$_0$ for perfect fluid distribution satisfying specific equation of state. Shri Ram [20] presented an algorithm for generating exact perfect fluid solution of Einstein field equation not satisfying the equation of

---

[*]Correspondence: Jaipal Singh, Dept. of Math. & Statistics, Univ. College of Sci., MLS University, Udaipur- 313001, India
E-mail: jaipalsingh075@gmail.com

state, for spatially homogeneous cosmological model of Bianchi type VI$_0$. In Bianchi type VI$_0$ string cosmology Tikekar and Patel [24] obtained some exact solutions.

Bianchi type VI$_0$ magnetized barotropic bulk viscous fluid massive string universe in general relativity have been obtained by Bali, Pradhan and Amirhshchi [6]. Verma and Ram [27] investigated Bianchi type VI$_0$ bulk viscous fluid models with variable gravitational and cosmological constant. Asgar and Ansari [1] have investigate accelerating Bianchi type VI$_0$ bulk viscous cosmological models in lyra geometry.

In the early universe, all the investigated dealing with physical process use a model of the universe, usually called a big-bang model. The big-bang model based on Einstein field equation successfully explains the three important observation in astronomy : (i) The phenomena of expanding universe, (ii) primordial nucleo- synthesis, (iii) The observed isotropy of the cosmic background radiation. Thus alternative theories were proposed from time to time, the most well known being the steady state theory by Bondi and Gold [11].

In this approach the universe does not have any singular beginning nor an end on the cosmic time scale moreover the statistical properties of the large scale features of the universe do not change. To account for the constancy of the mass density they have envisage a very slow but continuous creation matter in contrast to the one time infinite and explosive creation at t = 0 of the standard model. But it suffers from the serious disqualification that they do not give any physical justification in the form of any dynamical theory for the phenomenon of the continuous creation of matter. Thus the principle of conservation of matter is clearly scarified in this formalism. The overcome this difficulty in particular and host offers Hoyle and Narlikar [16] in this sixties adopted a field theoretic approach introducing a massless and chargeless scalar field C in the Einstein - Hilbert action to account for the matter creation.

Bali and Tikekar [7] have investigated C-field cosmological model for dust distribution in FRW space time with variable gravitational constant. Bali[2] studied Aspects of Inflation in Spatially Homogeneous and Anisotropic Bianchi type – I Spacetime with Exponential potential. Bali and Kumawat [5] have investigated C-field cosmological models for dust and barotropic fluid distribution in non flat FRW space time with variable gravitational constant. Bali and Saraf [4] have investigated Bianchi type I dust field universe with decaying vacuum energy in C-field cosmology. Saraf [23] studied Bianchi type I cosmological model for dust distribution with variable G and Lambda.

Narlikar and Padmanabhan [18] have investigated the solution of Einstein's field equation which admit radiation and negative energy massless scalar C-field as source. Tyagi and Singh [25] studied LRS Bianchi type III barotropic fluid cosmological model in C-field with varying cosmological constant Λ. LRS Bianchi type-V perfect fluid cosmological model in C-field theory with variable Λ have studied by Tyagi and Singh [26]. Parikh and Tyagi [19] have investigated Bianchi type VI$_0$ cosmological with Barotropic perfect fluid in Creation field theory with time dependent Λ. Mehta and Chundawat [17] have studied LRS Bianchi type II cosmological model with barotropic perfect fluid in C –Field theory with time dependent term.

In this paper, we have investigated Bianchi type-VI$_0$ cosmological model with bulk viscosity and dust distribution in C-field theory. We assume the matter content of the universe is in the form of dust, which leads to p = 0. To get the deterministic model of the universe we assumed $A = B^2$, where A and B are metric potentials. We find that the creation field (C) increase with time which matches with the result of H-N theory. The physical and geometrical aspects of the model are also discussed.

2. The Metric and Field Equations

We consider Bianchi type-VI$_0$ space time in the form of

$$ds^2 = -dt^2 + A^2 dx^2 + B^2 e^{2x} dy^2 + C^2 e^{-2x} dz^2 \qquad \text{... (1)}$$

where A, B and C are function of t alone.

Hoyle and Narlikar [16] modified the Einstein's field equation by introducing C-field as :

$$R_i^j - \frac{1}{2} R g_i^j = -8\pi G \left[\underset{(m)}{T_i^j} + \underset{(c)}{T_i^j} \right] \qquad \text{... (2)}$$

The energy momentum tensor $\underset{(m)}{T_i^j}$ for a bulk viscous fluid distribution and for creation field $\underset{(c)}{T_i^j}$ are given by

$$\underset{(m)}{T_i^j} = (\overline{p} + \rho)\, v_i v^j + \overline{p} g_i^j \qquad \text{... (3)}$$

where $\overline{p} = p - \xi\theta$

$$\underset{(c)}{T_i^j} = -f\left(c_i c^j - \frac{1}{2} g_i^j c_\alpha c^\alpha \right) \qquad \text{... (4)}$$

where f > 0 is coupling constant between the matter and creation field and $C_i = \dfrac{dC}{dx^i}$

The co-moving coordinate are chosen such that $v^i = (0,0,0,1)$. The non-vanishing components of energy momentum tensor for matter are given by

$$\underset{(m)}{T_1^1} = -\xi\theta = T_2^2 = T_3^3 \ , \ \underset{(m)}{T_4^4} = -\rho \qquad\qquad \dots (5)$$

The non-vanishing components of energy momentum tensor for creation field are given by

$$\underset{(c)}{T_1^1} = -\frac{1}{2} f\dot{C}^2 = T_2^2 = T_3^3, \ \underset{(c)}{T_4^4} = \frac{1}{2} f\dot{C}^2 \qquad\qquad \dots (6)$$

Hence the Einstein field equation (2) for the metric (1) and energy momentum tensor (5) and (6) takes the form

$$\frac{B_{44}}{B} + \frac{C_{44}}{C} + \frac{B_4 C_4}{BC} + \frac{1}{A^2} = 8\pi G\left[\xi\theta + \frac{1}{2} f\dot{C}^2\right] \qquad\qquad \dots (7)$$

$$\frac{A_{44}}{A} + \frac{C_{44}}{C^2} + \frac{A_4 C_4}{AC} - \frac{1}{A^2} = 8\pi G\left[\xi\theta + \frac{1}{2} f\dot{C}^2\right] \qquad\qquad \dots (8)$$

$$\frac{A_{44}}{A^2} + \frac{B_{44}}{B} + \frac{A_4 B_4}{AB} - \frac{1}{A^2} = 8\pi G\left[\xi\theta + \frac{1}{2} f\dot{C}^2\right] \qquad\qquad \dots (9)$$

$$\frac{A_4 B_4}{AB} + \frac{B_4 C_4}{BC} + \frac{A_4 C_4}{AC} - \frac{1}{A^2} = 8\pi G\left[\rho - \frac{1}{2} f\dot{C}^2\right] \qquad\qquad \dots (10)$$

$$\frac{1}{A}\left[\frac{C_4}{C} - \frac{B_4}{B}\right] = 0 \qquad\qquad \dots (11)$$

The suffix 4 by the symbols A, B and C denotes differentiation w.r.t. 't'.

3. Solutions of Field Equations

The conservation equation

$$[8\pi G T_i^j]_{;j} = 0 \qquad\qquad \dots (12)$$

which leads to

$$8\pi G\left[\dot{\rho} - f\dot{C}\ddot{C} + \{(\rho + \bar{p}) - f\dot{C}^2\}\left(\frac{A_4}{A} + \frac{B_4}{B} + \frac{C_4}{C}\right)\right] = 0 \qquad\qquad \dots (13)$$

following Hoyle and Narlikar [16], the source equation of C-field $C_{;i}^{i} = \dfrac{n}{f}$ leads to $C = t$ for large r, thus $\dot{C} = 1$.

Equation (11) leads to

$$C = l \; B \qquad\qquad \text{... (14)}$$

where l is constant of integration. Without loss of generality we assume $l = 1$.

To get deterministic solution of equation (7) to (10), we assume a condition between the metric potential as

$$A = B^2 \qquad\qquad \text{... (15)}$$

Equation (7) - (10) after using (14) and (15) leads to

$$\frac{2B_{44}}{B} + \frac{B_4^2}{B^2} + \frac{1}{B^4} = 8\pi G \left[\xi\theta + \frac{1}{2}f\dot{C}^2 \right] \qquad\qquad \text{... (16)}$$

$$\frac{3B_{44}}{B} + \frac{4B_4^2}{B^2} - \frac{1}{B^4} = 8\pi G \left[\xi\theta + \frac{1}{2}f\dot{C}^2 \right] \qquad\qquad \text{... (17)}$$

$$\frac{3B_{44}}{B} + \frac{4B_4^2}{B^2} - \frac{1}{B^4} = 8\pi G \left[\xi\theta + \frac{1}{2}f\dot{C}^2 \right] \qquad\qquad \text{... (18)}$$

$$\frac{5B_4^2}{B^2} - \frac{1}{B^4} = 8\pi G \left[\rho - \frac{1}{2}f\dot{C}^2 \right] \qquad\qquad \text{... (19)}$$

Equation (16) and (17) together leads to (Using $\dot{C}^2 = 1$)

$$2B_{44} + \frac{6B_4^2}{B} = \frac{4}{B^3} \qquad\qquad \text{... (20)}$$

Let $B_4 = f(B)$ which leads to $B_{44} = ff'$

Equation (20) leads to

$$\frac{df^2}{dB} + \frac{6}{B}f^2 = \frac{4}{B^3} \qquad\qquad \text{... (21)}$$

Equation (21) leads to

$$f^2 = \frac{1}{B^2} \qquad \qquad \text{... (22)}$$

The constant of integration has been taken zero for simplicity.

Equation (22) leads to

$$BdB = dt \qquad \qquad \text{... (23)}$$

Equation (23) leads to

$$B = (2t + \alpha)^{1/2} \qquad \qquad \text{... (24)}$$

Also the metric (1) reduces to

$$ds^2 = -dt^2 + (2t + \alpha)^2 [dx^2 + e^{2x} dy^2 + e^{-2x} dz^2] \qquad \qquad \text{... (25)}$$

Equation (14) and (15) in equation (13) we have

$$8\pi G \left[\dot{\rho} - fC\ddot{C} + \{(\rho - \xi\theta) - f\dot{C}^2\}\left(\frac{4B_4}{B}\right) \right] = 0 \qquad \qquad \text{... (26)}$$

Equation (26) leads to

$$\frac{d\dot{C}^2}{dt} + \frac{8}{(2t + \alpha)}\dot{C}^2 = \frac{8}{(2t + \alpha)} \qquad \qquad \text{... (27)}$$

Equation (27) leads to

$$\dot{C}^2 (2t + \alpha)^2 = (2t + \alpha)^2 + Q \qquad \qquad \text{...(28)}$$

For simplicity, we assume that integral constant $Q = 0$ which leads to

$$\dot{C}^2 = 1 \qquad \qquad \text{... (29)}$$

Thus, we have $\dot{C} = 1$ or $C = t$ 　　　　... (30)

We find $C = t$, which agrees with the value used in the source equation. Thus creation field C is proportional to time t.

ISSN: 2153-8301　　　　　　　　　Prespacetime Journal　　　　　　www.prespacetime.com
Published by QuantumDream, Inc.

4. Some Physical and Geometrical Features

The homogeneous mass density (ρ), the creation field (C), spatial volume (R^3), the deceleration parameter (q), shear tensor (σ) and expansion (θ) of the model (25) are given by

$$\rho = \frac{1}{8\pi G}\left[\frac{4}{(2t+\alpha)^2}\right] + \frac{f}{2} \qquad \text{... (31)}$$

$$C = t \qquad \text{... (32)}$$

$$R^3 = ABC = (2t+\alpha)^2 \qquad \text{... (33)}$$

$$q = (-)\frac{\ddot{R}/R}{\dot{R}^2/R^2} = 1 \qquad \text{... (34)}$$

$$\theta = \frac{4}{(2t+\alpha)} \qquad \text{... (35)}$$

$$\sigma^2 = \frac{1}{3}\left(\frac{1}{(2t+\alpha)^2}\right) \qquad \text{... (36)}$$

$$\frac{\sigma}{\theta} = \text{Constant} \neq 0 \qquad \text{... (37)}$$

5. Conclusion

The scale factor R increase with time, since deceleration parameter $q > 0$, hence the model (25) represent decelerating universe. The homogeneous density decrease as time increase. The model (25) passes through a singular state at $t = -\alpha/2$, this is explained as Creation exists all the time , so there is a big crunch between $-\alpha/2$ to ∞ and Creation is going on front $t = -\alpha/2$ to ∞. During this period the model exist. Since $t \to \infty, \frac{\sigma}{\theta}$ is constant, therefore model (25) does not approaches to isotropy in late time. Since $\theta \neq \infty$ at $t = 0$, hence model (25) is free from initial singularity.   The coordinate distance γ_H to the horizon is the maximum distance a null ray could have travelled at time t starting from infinite past i.e.

$$\gamma_H = \int_{-\infty}^{t}\frac{dt}{R^3(t)}$$

We could extent the proper time t to in the past because of non-singular nature of space time, thus

$$\gamma_H(t) = \int_0^t \frac{dt}{R^3(t)} = \int_0^t \frac{dt}{(2t+\alpha)^2}$$

The integral of diverges at lower limit shows that the model is free from event horizon.

Received November 13, 2020; Accepted December 21, 2020

References

1. Asgar, A. and Ansari, M. (2014). Accelerating Bianchi type-VI₀ bulk viscous cosmological model in Lyra geometry, J. Theor. Appl. Phys., 8, 219-224.

2. Bali, R.(2018). Aspects of Inflation in Spatially Homogeneous and Anisotropic Bianchi type – I Spacetime with Exponential potential, Modern Phys. Lett. A, 33(4),1850238-1850245.

3. Bali, R. and Goyal, R. (2019).Inflationary Scenario in Bianchi type- V Spacetime with Variable Bulk Viscosity and Dark Energy in Radiation Dominated Phase, Prespacetime Journal, 10(1), 68- 77.

4. Bali, R. and Saraf, S. (2012). Bianchi type-I dust field universe with decaying vacuum energy in C-field cosmology, IJRRAS, 13(3), 800-805.

5. Bali, R. and Kumawat, M. (2009). C-field cosmological model with variable G in FREW space time, Int. J. Theor. Phys., 48, 3410.

6. Bali, R., Pradhan, A. and Amirhas Chi, H. (2008). Bianchi type-VI₀ magnetized barotropic bulk viscous fluid massive string universe in general relativity, Int. J. Theor. Phys., 47, 2594-2604.

7. Bali, R. and Tikekar, R. (2007). C-field cosmology with variable G in flat FRW model, Chin. Phys. Lett., 24(11), 2390.

8. Bali, R. and Singh, D.K. (2005). Bianchi Type-V bulk viscous fluid string dust cosmological model in general relativity, Astrophsa. Spae Science, 300, 387-394.

9. Bali, R. (1986). The behaviour of the magnetic field in a cosmological model for perfect fluid distribution, Int. J. Theor. Phys. 25, 7.

10. Barrow, J.D. (1984). Heliam formation in cosmologies with Anisotropic Curvature, Royal Astronomical Society, 211(1), 221-227.

11. Bondi, H. and Gold, J. (1948). The steady state theory of the expanding universe, Monthly Notices of the Royal Astronomical Society, 108, 252.

12. Collins, C.B. and Hawking, S.W. (1973). Why is the universe. Isotropic, Astrophys. J., 180, 317-334.

13. Collins, C.B. (1971). More qualitative cosmology, Communications in Mathematical Physics, 23(2), 137-158.

14. Dum, K. and Tupper, B.O.J. (1978). Tilting and viscous models in a class of type VI₀ cosmologies, Astrophys. J., 222, 405-411.

15. Ellis, G.F.R. and McCollum, M.A.H. (1969). A class of homogeneous cosmological models, Communications in Mathematical Physics, 12(2), 108-141.

16. Hoyle, F. and Narlikar, J.V. (1964). A new theory of gravitation, IUCAA, Proc. Roy. Soc., 282, 191-207.

17. Mehta, P. and Chundawat, P.K. (2020). LRS Bianchi type II cosmological with Barotropic Perfect fluid in C – field theory with time dependent term, International Journal Of Scientific Development and Research, 5(1), 140-145.

18. Narlikar, J.V. and Padmanabhan, J. (1985). Creation field cosmology : A possible solution to singularity, Horizon and Flatness Problems, Phys. Rev. D., 32, 1928.

19. Parikh, S. and Tyagi, A. (2016). Time dependent Λ in Bianchi type IX space time with Barotropic perfect fluid in C-field theory, Prespace Journal, 7(12), 1645.

20. Ram, S. (1989). LRS Bianchi-I perfect fluid solutions generated from known solutions, International Journal of Theoretical Physics, 28(8), 917-921.

21. Roy, S.R. and Bali, R. (1984). Some magnetohydrostatic models of cylindrical symmetry in general relativity, J. Math. Phys., 25, 1456.

22. Ruban, V.A. (1978). Lennigral Institute of Nuclear Phys., BP Konstantinova, Preprint, 411.

23. Saraf, T. (2017). Bianchi type I cosmological model for dust distribution with variable G and Λ, JRAOPS,16, 175-182.

24. Tikekar, R. and Patel, L.K. (1994). Some exact solutions in Bianchi type VI$_0$ string cosmology, Pramana, 42(6), 483-489.

25. Tyagi, A. and Singh, G.P. (2015). Time dependent Λ in C-field theory with LRS Bianchi type-III universe and barotropic perfect fluid, Prespace time Journal, 6(2), 143-150.

26. Tyagi, A. and Singh, G. (2015). LRS Bianchi type-V perfect fluid cosmological model in C-field theory with variable Λ, Journal of Chemical, Biological and Physical Sciences, 5(2), 1878-1883.

27. Verma, M.K. and Ram, S. (2011). Bianchi type VI$_0$ bulk viscous fluid models with variable gravitational and cosmological constants, Applied Mathematics, 2(03), 348.

Article

Some Characterizations on the Special Tubular Surfaces in Galilean Space

Fatma Almaz[*] & Mihriban Alyamaç Külahci

Dept. of Math., Faculty of Science, Firat University, Elaziğ, Turkey

Abstract

In this paper, the tubular surface generated by rectifying curves are explored. Also, using the Gaussian and mean curvatures of tubular surfaces, for the linear Weingarten surfaces and HK-quadric surfaces, harmonic surfaces some characterizations are given.

Keywords: Galilean space, tubular surfaces, rectifying curves, Weingarten surface, HK-quadric surface.

1. Introduction

In mathematics, a surface is a geometrical shape that corresponds a disfigured plane. The most common examples arise as boundaries of solid objects in ordinary three-dimensional Euclidean space. In the study of the differential geometry of surfaces, it is common to determine some surfaces satisfying curvature conditions. Also, the differential geometry of surfaces deals with the differential geometry of smooth surfaces with various additional structures for example a Riemannian metric. Surfaces have been extensively studied from various perspectives: extrinsically, relating to their embedding in Euclidean space and intrinsically, reflecting their properties determined solely by the distance within the surface as measured along curves on the surface.

A surface whose mean curvature is in functional relationship with its Gaussian curvature. Namely, a surface is said to be a Weingarten surface if there exists a relation, that does not depend on the parameters, between the mean curvature and the total curvature (or between the principal curvatures). Also, a surface is said to be a Weingarten if there is a smooth relation $U(k_1, k_2) = 0$ between two principle curvatures k_1 and k_2. If K and H denote the Gauss curvature and the mean curvatures, respectively, then $U(k_1, k_2) = 0$ implies a relation as $\Phi(K, H) = 0$. The existence of a non-trivial functional relation $\Phi(K, H) = 0$ on a surface, which is parameterized by a patch $x(w, v)$, is equivalent to the following Jacobian determinant [9],

$$\frac{\partial(K,H)}{\partial(w,v)} = 0, \tag{1.1}$$

---

[*]Correspondence: V. M. Raut, Dept. of Math., Faculty of Science, Firat University, Elaziğ, Turkey.
  E-mail: fb_fat_almaz@hotmail.com

Furthermore, if the equations $U = a_1 k_1 + a_2 k_2 - a_3$ or $\Phi = a_1 H + a_2 K - a_3$ hold, the surfaces are called linear Weingarten surfaces, where $a_i, i \in \mathbb{R}$ with $a_1^2 + a_2^2 \neq 0$. The following points are important for consideration:

1. If the constant $a_2 = 0$, a linear Weingarten surface reduces to a surface with constant Gaussian curvature.

2. If the constant $a_1 = 0$, a linear Weingarten surface reduces to a surface with constant mean curvature.

On the other hand, if a surface satisfies the following equation

$$a_1 H^2 + 2a_2 HK + a_3 K^2 = \text{constant}, a_1 \neq 0, \tag{1.2}$$

then the surface is said to be a HK −quadric surface, [6]. Also, the surfaces of revolution and surfaces with constant mean or constant Gaussian curvature are given as examples of Weingarten surfaces. Also, the linear Weingarten surfaces can be expressed as natural generalization of surfaces with constant Gaussian curvature or constant mean curvature, [9].

In [1], timelike tube surface around the spacelike curve with timelike and spacelike binormal vectors is studied in a three-dimensional Minkowski space E_1^3 by the authors. Moreover, Weingarten and linear Weingarten conditions are given for this surface with respect to their curvatures. The same studies and consequences about surfaces in G_3 are given by the authors in [2, 3]. In [8], the surfaces in Euclidean 3-space foliated by pieces of circles are studied and that satisfy a Weingarten condition of type $aH + bK = c$; $a, b, c \in \mathbb{R}$, H and K denote the mean curvature and the Gauss curvature respectively, by the author. In [12], a tube in a Euclidean 3-space satisfying some equation in terms of the Gaussian curvature, the mean curvature and the second Gaussian curvature are studied by the authors.

In [14], that parallel surfaces of a non-developable ruled surface are not ruled surfaces by using fundamental forms is studied. Also, that parallel surfaces of ruled Weingarten surface are Weingarten surface is shown by the authors. Furthermore, ruled Weingarten surfaces in the Galilean space are studied by the authors, in [15]. Weingarten surfaces are surfaces having a nontrivial funcional relation between their Gaussian and mean curvature. The same consequences about surfaces and curves in G_3 and pseudo Galilean space are given by the authors in [9, 10, 17]. Also, the some consequences about surfaces and curves in different ambiant spaces are investigated by the authors in [5, 6, 16].

2. Preliminaries

Classical context of the Euclidean space is the origin of results, which could be transferred to some other geometries. One way of defining new geometries is through Cayley-Klein spaces. They are expressed as projective spaces, $P_n R$, with an absolute figure, which is a subset of $P_n R$ originating as sequence of quadrics and planes 1. The projective space, $P_n R$, has invariants as the absolute figure definitition for the subgroup of projectivities named as the Cayley-Klein space group of movements. By means of the absolute figure, metric connections are defined and they are invariant under the group of movements.

The scalar product of the vectors $U = (u_1, u_2, u_3)$, $V = (v_1, v_2, v_3)$ in G_3 is defined as,

$$\langle U, V \rangle = \begin{cases} u_1 v_1, & \text{if } u_1 \neq 0 \text{ or } v_1 \neq 0 \\ u_2 v_2 + u_3 v_3, & \text{if } u_1 = 0, v_1 = 0 \end{cases} \tag{2.1}$$

The cross product in Galilean space is given as

$$U \times V = \begin{cases} (0, v_1 u_3 - v_3 u_1, v_2 u_1 - v_1 u_2), & \text{if } u_1 \neq 0 \text{ or } v_1 \neq 0 \\ (v_3 u_2 - v_2 u_3, 0, 0), & \text{if } u_1 = 0, v_1 = 0 \end{cases} \tag{2.2}$$

Let $\beta: I \subset \mathbb{R} \to G_3$ be a curve given by $\beta(s) = (s, y(s), z(s))$. The vectors of the Frenet-Serret frame are defined by

$$t(s) = \beta'(s) = (1, y'(s), z'(s)); n(s) = \frac{t'(s)}{\kappa(s)}; b(s) = \frac{n'(s)}{\tau(s)},$$

where the real valued function $\kappa(s) = \| t'(s) \|$ is given as the first curvature of the curve β, the second curvature function is defined as $\tau(s) = \| n'(s) \|$. For the curve in G_3, Frenet-Serret equations are written as follows

$$t' = \kappa n, n' = \tau b, b' = -\tau n. \tag{2.3}$$

Let the equation of a surface $\Theta = \Theta(s, v)$ in G_3 be given by
$$\Theta(s, v) = (x(s, v), y(s, v), z(s, v)). \tag{2.4}$$

Then the unit isotropic normal vector field η on $\Theta(s, v)$ becomes as

$$\eta = \frac{\Theta_{,1} \times \Theta_{,2}}{\| \Theta_{,1} \times \Theta_{,2} \|}, \tag{2.5}$$

where the partial differentiations with respect to s and v will be denoted as follows

$$\Theta_{,1} = \frac{\partial \Theta(s,v)}{\partial s}; \Theta_{,2} = \frac{\partial \Theta(s,v)}{\partial v}. \tag{2.6}$$

On the other hand, the isotropic unit vector δ on the tangent plane of the surface is given by

$$\delta = \frac{x_{,2}\Theta_{,1} - x_{,1}\Theta_{,2}}{w}, \tag{2.7}$$

where $x_{,1} = \frac{\partial x(s,v)}{\partial s}$, $x_{,2} = \frac{\partial x(s,v)}{\partial v}$ and $w = \| \Theta_{,1} \times \Theta_{,2} \|$.

Let us define

$$g_1 = x_{,1}, g_2 = x_{,2}, g_{ij} = g_i g_j; g^1 = \frac{x_{,2}}{w}; g^2 = \frac{x_{,1}}{w}; g^{ij} = g^i g^j; i, j = 1,2 \tag{2.8}$$

$$h_{11} = \langle \Theta_{,1}^*, \Theta_{,1}^* \rangle, h_{12} = \langle \Theta_{,1}^*, \Theta_{,2}^* \rangle; h_{22} = \langle \Theta_{,2}^*, \Theta_{,2}^* \rangle, \tag{2.9}$$

where $\Theta_{,1}^*$ and $\Theta_{,2}^*$ are the projections of the vectors $\Theta_{,1}$ and $\Theta_{,2}$ onto the yz-plane, respectively. The first fundamental form ds^2 of the surface $\Theta(s, v)$ is given as, [7, 11],

$$ds^2 = ds_1^2 + ds_2^2 = (g_1 ds + g_2 dv)^2 + \varepsilon(h_{11}ds^2 + 2h_{12}dsdv + h_{22}dv^2), \tag{2.10}$$

where

$$\varepsilon = \begin{cases} 0, & dw:dv_1 \text{non} - \text{isotropic} \\ 1, & dw:dv_1 \text{isotropic} \end{cases} \tag{2.11}$$

In this case, the coefficients of ds^2 are denoted by g_{ij}^*. The function can be represented in terms of g_i and h_{ij} as follows

$$w^2 = g_1^2 h_{22} - 2g_1 g_2 h_{12} + g_2^2 h_{11}.$$

The Gaussian curvature and the mean curvature of a surface are defined by means of the second fundamental form L_{ij} coefficients, which are the normal components of $\Theta_{,i,j}(i,j = 1,2)$. Namely,

$$\Theta_{,i,j} = \sum_{k=1}^{2} \Gamma_{ij}^k \Theta_{,k} + L_{ij}\eta, \tag{2.12}$$

where Γ_{ij}^k is the Christoffel symbols of the surface and L_{ij} are given as

$$L_{ij} = \frac{1}{g_1}\langle g_1 \Theta_{,i,j}^* - g_{i,j}\Theta_{,1}^*, \eta \rangle = \frac{1}{g_2}\langle g_2 \Theta_{,i,j}^* - g_{i,j}\Theta_{,2}^*, \eta \rangle. \tag{2.13}$$

From this, the Gaussian curvature K and the mean curvature H of the surface are given as,

$$K = \frac{L_{11}L_{22} - L_{12}^2}{w^2}, H = \frac{g_2^2 L_{11} - 2g_1 g_2 L_{12} + g_1^2 L_{22}}{w^2}, \tag{2.14}$$

[7, 11, 13].

Definition 1 *A vector $x = (x_1, x_2, x_3)$ is called a non-isotropic if $x_1 \neq 0$. All unit isotropic vectors are of the form $x = (1, x_2, x_3)$. For isotropic vectors, $x_1 = 0$ hold, [7].*

Proposition 1 *Δf Laplacian of the differentiable function given by $f:U \subset \mathbb{R}^2 \to \mathbb{R}$ is defined as*

$$\Delta f = \left(\frac{d^2 f}{du^2}\right) + \left(\frac{d^2 f}{dv^2}\right), (u,v) \in U.$$

If $\Delta f = 0$ then the function f is harmonic in U, [4].

3. The Special Tubular Surfaces Generated by Rectifying Curves in Galilean 3-Space

In this work, the tubular surface generated by rectifying curves are examined, and using the Gaussian and mean curvatures of the special tubular surfaces, the conditions being linear, Weingarten surfaces and HK −quadric surface are expressed.

3.1   Characterization of isotropic rectifying curves in G_3

In this subsection, using the position components of vectors' curvature functions, the rectifying curves in G_3 can be described.

Theorem 1 *Let $\beta:I \subset \mathbb{R} \to G_3$ be a regular isotropic curve with curvatures $\kappa(w) \geq 0, \tau$ in G_3. Then, β is a rectifying curve if and only if the position vector of β satisfies the vector equation*

$$\beta(w) = (w + c)\overleftarrow{t} + \frac{\kappa(w)(w+c)}{\tau}\overleftarrow{b} \ and \ \tau(w) = \frac{\kappa(w)(w+c)}{d}, c, d \in \mathbb{R}_0. \qquad (3.1)$$

Proof. Assume that $\beta(w)$ is an rectifying curve with the curvature functions $\kappa(w)$, $\tau(w)$ in G_3 as follows

$$\beta(w) = \Sigma_0 \overleftarrow{t} + \Sigma_1 \overleftarrow{b}, \qquad (3.2)$$

for some differentiable functions $\Sigma_0(w)$, $\Sigma_1(w)$ and differentiating (3.2) with respect to w and using (2.3), one can obtain

$$\overleftarrow{t} = \Sigma_0' \overleftarrow{t} + (\Sigma_0 \kappa - \Sigma_1 \tau)\overleftarrow{n} + \Sigma_1' \overleftarrow{b}. \qquad (3.3)$$

Exposing the inner product t, n, b of the both side in (3.3), respectively, one can have

$$\Sigma_0' = 1; \Sigma_0 \kappa - \Sigma_1 \tau = 0; \Sigma_1' = 0. \qquad (3.4)$$

Using (3.4) and making necessary calculations, one may get,

$$\Sigma_0 = c + w; \Sigma_1 = dor\Sigma_1 = \frac{\kappa(w)(w+c)}{\tau}. \qquad (3.5)$$

Thus, one can find the position vector as,

$$\beta(w) = (c + w)\overleftarrow{t} + d\overleftarrow{b} = (w + c)\overleftarrow{t} + \frac{\kappa(w)(w+c)}{\tau}\overleftarrow{b}.$$

3.2   The mathematical approach on tube surfaces with rectifying curve in G_3

In this section, the tubular surfaces generated by rectifying curve are investigated according to mathematical approach. A canal surface is expressed as the envelope of a setting out sphere with exchanging radius, which is described by the orbit $\beta(w(s))$ (spine curve) with its center and a radius function ρ in addition to its parametrized through Frenet frame of the spine curve $\beta(w(s))$. If the radius function ρ is a constant, then the canal surface is called as a tube. Let one denotes by ρ the vector connecting the point from the parametrized curve $\beta(w(s))$ with the point from the surface, one can have the position vector R of a point on the surface as

$$R = \beta(w(s)) + \rho, i \in \mathbb{N}, \qquad (3.6)$$

and since ρ lies in the Euclidean normal plane of the curve $\beta(w(s))$, the points at a distance A_1 from a point of $\beta(w(s))$ form an Euclidean circle in G_3, [3]. Thus, it can be written as $\rho = A_1(\cos v_1 \overrightarrow{n} + \sin v_1 \overrightarrow{b})$, where v_1 is the Euclidean angle between the isotropic vectors $\overleftarrow{n}$ and $\overrightarrow{p}$.

Let $\Theta(w, v_1)$ be the tube surface generated by rectifying curve and let $\beta: I \subset \mathbb{R} \to G_3$ be a regular isotropic curve with curvatures $\kappa(w) \geq 0, \tau$ in G_3. Then, the tube surface generated by rectifying curve is parametrized as

$$\Theta(w, v_1) = \beta(w(s)) + A_1(\cos v_1(s)\overleftarrow{n} + \sin v_1(s)\overleftarrow{b}), \qquad (3.7)$$

where angle v_1 lies between the isotropic vectors $\overleftarrow{n}$ and $\overleftarrow{R} = A_1$. Clearly, one can get,

$$\Theta(w, v_1) = (w + c)\overleftarrow{t} + A_1\cos v_1\overleftarrow{n} + (d + A_1\sin v_1)\overleftarrow{b} \tag{3.8}$$

or

$$\Theta(w, v_1) = (w + c)\overleftarrow{t} + A_1\cos v_1\overleftarrow{n} + (\frac{\kappa(w)(w+c)}{\tau} + A_1\sin v_1)\overleftarrow{b}. \tag{3.9}$$

Then, one can get partial derivatives of $\Theta(w, v_1)$ with respect to w and v_1 as follows

$$\Theta_w = \overleftarrow{t} + \left((w + c)\kappa - \tau(d + A_1\sin v_1)\right)\overleftarrow{n} + \tau A_1\cos v_1\overleftarrow{b} = N_w, \tag{3.10}$$
$$\Theta_{v_1} = A_1(-\sin v_1\overleftarrow{n} + \cos v_1\overleftarrow{b}) = A_1 N_{v_1}; \tag{3.11}$$

It follows that the vector cross product of them is obtained as

$$\Theta_w \times \Theta_{v_1} = -A_1\cos v_1\overleftarrow{n} - A_1\sin v_1\overleftarrow{b}; \tag{3.12}$$
$$\left\|\Theta_w \times \Theta_{v_1}\right\| = A_1. \tag{3.13}$$

From previous equations, by using (3.12) and (3.13), the unit isotropic normal vector η of $\Theta(w, v_1)$ is given as follows $\eta = -\cos v_1\overleftarrow{n} - \sin v_1\overleftarrow{b}$. Furthermore, from (2.7), one can obtain $\delta = \frac{-\Omega_{v_1}^{\overleftarrow{1}}}{A_1} = \sin v_1\overleftarrow{n} - \cos v_1\overleftarrow{b}$. Since, $\overleftarrow{n}$ and $\overleftarrow{b}$ are the isotropic vectors, the Galilean frenet frame, usage leads to,

$$x(w, v_1) = w + c; x_w = 1 = g_1; x_{v_1} = 1 = g_2; g_{11} = 1, g_{12} = 0, g_{22} = 0; \tag{3.14}$$
$$g^1 = 0, g^2 = \frac{-1}{A_1}; h_{11} = 1, h_{12} = 0, h_{22} = A_1^2. \tag{3.15}$$

After the substitution of (3.14) and (3.15) into (2.10), the coefficients of the first fundamental form of the tubular surface can be obtained with the Galilean frenet frame in Galilean space as

$$I = dw^2 + \varepsilon(dw^2 + A_1^2 dv_1^2) or I = 2dw^2 + A_1^2 dv_1^2; \varepsilon = 1. \tag{3.16}$$

If one wants to calculate the second fundamental form of $\Phi(w, v_1)$, it is then necessary have to compute the following equations

$$\Theta_{ww} = (2\kappa + (w + c)\kappa' - \tau'(d + A_1\sin v_1) - \tau^2 A_1\cos v_1)\overleftarrow{n}$$
$$+(\tau\kappa(w + c) - \tau^2(d + A_1\sin v_1) + \tau'A_1\cos v_1)\overleftarrow{b},$$
$$\Theta_{v_1 v_1} = A_1(-\cos v_1\overleftarrow{n} - \sin v_1\overleftarrow{b}); \Theta_{wv_1} = -\tau A_1\cos v_1\overleftarrow{n} - \tau A_1\sin v_1\overleftarrow{b}. \tag{3.17}$$

The coefficients of the second fundamental form are calculated from (2.13) and (3.14), (3.17), as follows

$$L_{11} = \left(-2\kappa(w) + \tau'd - \kappa'(w)(w + c)\right)\cos v_1 + \tau^2 A_1;$$
$$L_{22} = A_1; L_{12} = \tau A_1. \tag{3.18}$$

Thus, the Gaussian curvature K and the mean curvature H are expressed as

$$K = \frac{-\kappa(w)\cos v_1}{A_1}(\tau d = \kappa.(w+c)), \qquad (3.19)$$

$$H = \frac{1}{2A_1}. \qquad (3.20)$$

and from $\tau(w) = \frac{(w+c)\kappa(w)}{d}$ and (3.19), (3.20), the curvatures of the rectifying curve are obtain as

$$\kappa(w) = \frac{-K}{2H\cos v_1}; \tau(w) = \frac{-(w+c)K}{2dH\cos v_1}.$$

Hence, the following theorem can be given.

Theorem 2 *Let* Θ *be a tubular surface by generated a rectifying curve in* G_3. *Then, the curvatures of the rectifying curve are given as*

$$\kappa(w) = \frac{-K}{2H\cos v_1}; \tau(w) = \frac{-(w+c)K}{2dH\cos v_1}.$$

Theorem 3 *A tubular surface* Θ *by generated a rectifying curve in* G_3 *is also a* $\Phi(K,H)$-*Weingarten surface.*

Proof. From definition of the Weingarten surface, taking derivative of K and H with respect to w and v_1 yields,

$$K_w = \frac{(\tau_{ww}(w)d - \kappa_{ww}(w)(w+c) - 3\kappa_w(w))\cos v_1}{A_1}, H_w = 0;$$

$$K_{v_1} = \frac{-(\tau_w(w)d - (w+c)\kappa_w(w) - 2\kappa(w))\sin v_1}{A_1}, H_{v_1} = 0.$$

Since $\frac{(w+c)\kappa(w)}{d} = \tau(w)$, for a rectifying curve, one can write

$$K_w = \frac{-\kappa_w(w)\cos v_1}{A_1}, H_w = 0; K_{v_1} = \frac{\kappa(w)\sin v_1}{A_1}, H_{v_1} = 0. \qquad (3.21)$$

Furthermore, if the tubular surface Θ generated by a rectifying curve in G_3 satisfies the equation $\Phi(X,Y) = 0$, then the surface is called as $\Phi(K,H)$-Weingarten surface. Therefore, by using (3.21), we get

$$\Phi(K,H) = \frac{\partial(K,H)}{\partial(w,v_1)} = K_w H_{v_1} - K_{v_1} H_w = 0,$$

and we say that the surface Θ is a $\Phi(K,H)$-Weingarten surface.

Theorem 4 *If* Θ *is a linear Weingarten surface in* G_3, *for* $a_1 \neq 0, v_1 \neq \frac{(2n+1)\pi}{2}, n \in \mathbb{N}$, *the tubular surface* Θ *generated by the rectifying curve is a linear Weingarten surface, while it is also a flat surface reduced to a cylindrical surface with constant Gaussian curvature.*

Proof. Let Θ be a linear Weingarten surface in G_3, then from the definition of the Weingarten surface, by using the equations $K = \frac{-\kappa(w)\cos v_1}{A_1}, H = \frac{1}{2A_1}$, one can reach to

$$a_1 K + a_2 H = a_3$$

$$a_1 \cdot \frac{-\kappa(w)\cos v_1}{A_1} + a_2 \cdot \frac{1}{2A_1} = a_3$$

$$-\kappa(w)2a_1 \cdot \cos v_1 + (a_2 - a_3 2A_1) = 0.$$

From the definition of the linear independent of vectors,

$$-\kappa(w)2a_1 \cdot \cos v_1 = 0 \text{ and } (a_1 - a_3 2A_1) = 0$$

and by using previous equations, from $A_1 = \frac{-\kappa(w)\cos v_1}{K}$, one have

$$\frac{a_2}{2a_3} = A_1 ; \kappa(w) = 0 \Rightarrow \frac{a_2}{2a_3} = \frac{-\kappa(w)\cos v_1}{K} \text{ and } a_2 = 0. \tag{3.22}$$

Hence, the surface Θ is a cylinder and for the constant $a_2 = 0$, the surface Θ is reduced to a cylinder surface with constant Gaussian curvature. Therefore, from $\kappa(w) = 0$ and $K = \frac{-\kappa(w)\cos v}{A}$ equations, $K = 0$ is found that the surface is flat.

Theorem 5 *Let Θ be a tube surface with rectifying curve in G_3. For the parameter*

$$v_1 = arccos\left[\frac{a_2/2a_3}{\kappa(w)}\right] = arccos\left[\left(\frac{a_2}{a_3}\right)\frac{H}{K}\cos v_1\right],$$

the surface Θ is a $HK -$quadric surface.

Proof. Assume that the tubular surface Θ is $HK -$quadric surface. From definition of the $HK -$quadric surface, after taking necessary differentials it is possible to have calculations, we get

$$a_1 HH_w + a_2(H_w K + HK_w) + a_3 KK_w = 0.$$

Then, by using the equations $K_w = \frac{(\tau_{ww}(w)d - \kappa_{ww}(w)(w+c) - 3\kappa_w(w))\cos v_1}{A_1}$ and $H_w = 0$, one can obtain

$$a_2 HK_w + a_3 KK_w = 0,$$

$$(d\tau_{ww} - (w+c)\kappa_{ww} - 3\kappa_w)\cos v_1\big(a_2 + 2a_3\cos v_1(d\tau_w - (w+c)\kappa_w - 2\kappa)\big) = 0.$$

Later from the previous equation, one gets

$$(\tau_{ww}d - \kappa_{ww}(w+c) - 3\kappa_w) = 0$$

or

$$\big(a_2 + 2a_3\cos v_1(\tau_w d - \kappa_w(w+c) - 2\kappa)\big) = 0.$$

For a rectifying curve in G_3, since $(w+c)\kappa(w) = \tau(w)d$, one can get

$$(a_2 - \kappa 2a_3\cos v_1) = 0.$$

Hence, since $\kappa(w) = \dfrac{-K}{2Hcosv_1}$, the parameter v_1 can be written as

$$v_1 = arccos\left[\frac{a_2/2a_3}{\kappa(w)}\right] = arccos\left[\left(\frac{a_2}{a_3}\right)\frac{H}{K}cosv_1\right]. \tag{3.23}$$

Therefore, one can state that the tubular surface Θ generated by the rectifying curve is HK −quadric surface.

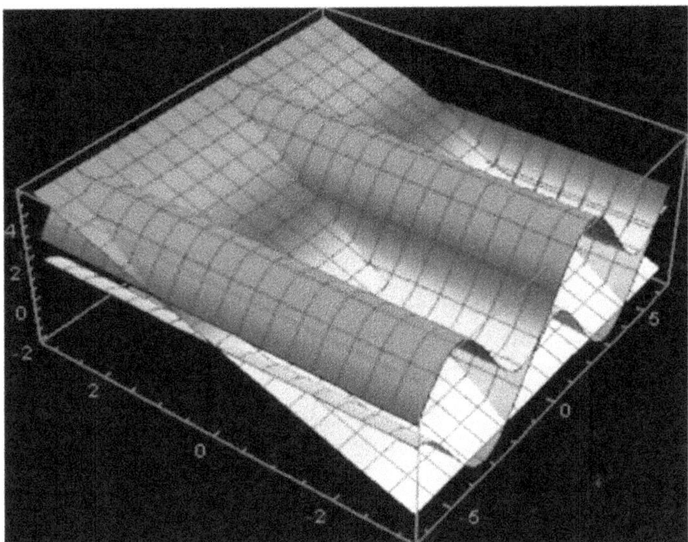

Figure 1. The Weingarten tube surface formed by the rectifying curve in Galilean space

Theorem 6 *The tubular surface Θ generated by a rectifying curve in G_3 is a harmonic $\Leftrightarrow$ the following expression is provided*

$$v = k\pi, k \in \mathbb{Z} \text{ or } v = \arccos\left(\frac{-1}{A}\frac{\partial^2\left(\frac{(w+c)\kappa(w)}{\tau(w)}\right)}{\partial w^2}\right).$$

Proof. Let the tube surface Θ be formed by the rectifying curve β in G_3. Also, in order for its Θ^i, $i = 1,2,3$ coordinates functions to be harmonic, it is necessary to provide $\Delta\Theta^i = 0$ equality from the definition. So, by making the necessary calculations in the following expression,

$$\Theta(w,v) = \left(w + c, Acosv, \frac{\kappa(w)(w+c)}{\tau(w)} + Asinv\right)$$
$$\Theta(w,v) = (\Theta^1, \Theta^2, \Theta^3),$$

the following equations can be written

$$\Delta\Theta^1_w = \frac{\partial^2\Theta^1}{\partial w^2} = 0, \Delta\Theta^1_v = \frac{\partial^2\Theta^1}{\partial v^2} = 0$$
$$\Delta\Theta^1 = 0;$$
$$\Delta\Theta^2_w = \frac{\partial^2\Theta^2}{\partial w^2} = 0, \Delta\Theta^2_v = \frac{\partial^2\Theta^2}{\partial v^2} = -Asinv$$
$$\Delta\Theta^2 = -Asinv = 0;$$

$$\Delta\Theta_w^3 = \frac{\partial^2\Theta^3}{\partial w^2} = \frac{\partial^2\left(\frac{(w+c)\kappa(w)}{\tau(w)}\right)}{\partial w^2}, \Delta\Theta_v^3 = \frac{\partial^2\Theta^3}{\partial v^2} = A\cos v$$

$$\Delta\Theta^3 = \frac{\partial^2\left(\frac{(w+c)\kappa(w)}{\tau(w)}\right)}{\partial w^2} + A\cos v = 0.$$

Hence, from the previous equations, respectively, one gets

$$-A\sin v = 0 \Rightarrow v = k\pi, k \in \mathbb{Z}$$

or

$$\frac{\partial^2\left(\frac{(w+c)\kappa(w)}{\tau(w)}\right)}{\partial w^2} + A\cos v = 0 \Rightarrow v = \arccos\left(\frac{-1}{A}\frac{\partial^2\left(\frac{(w+c)\kappa(w)}{\tau(w)}\right)}{\partial w^2}\right).$$

4. Conclusion

In this paper, the tube surfaces generated by rectifying curves are examined and some certain results according to the curvatures of the surfaces are presented in detail. Moreover, using the Gaussian and mean curvatures of tube surfaces generated by rectifying curve, it is possible to try and to express the conditions being linear Weingarten surfaces, $HK-$quadric surface and harmonic. The authors are currently working on the properties of these tubular and canal surfaces with a view to devising suitable metric in 3-Galilean and 4-Galilean spaces by adapting the type of conservation laws considered in the paper.

Received November 4, 2020; Accepted December 21, 2020

References

[1] Abdel-Aziz HS, Saad MK, Weingarten Timelike Tube Surfaces Around a Spacelike Curve, Int. J. Math. Anal. 2011; 5: 1225-1236.

[2] Ali AT, Position vectors of curves in the Galilean space G_3, Matematicki Vesnik. 2012; 64(3): 200-210.

[3] Dede M, Tubular surfaces in Galilean space, Math. Commun. 2013; 18(1): 209-217.

[4] Do Carmo, MP, Differential Geometry of Curves and Surface, Prentice-Hall, Inc. Englewood Cliffs, New Jersey, 1976.

[5] Karacan MK, Yayli Y, On the geodesics of tubular surfaces in Minkowski $3-$Space, Bull. Malays. Math. Sci. Soc. 2008; 31(1): 1-10.

[6] Kim YH, Yoon DW, On Non-Developable Ruled Surface in Lorentz Minkowski 3-Spaces, Taiwanese Journal of Mathematics, 2007; 11(1): 197-214.

[7] Kuhnel W, Differential Geometry Curves- Surfaces and Manifolds, Second Edition, Providence, RI, United States, American Math. Soc., 2006; 16, ISBN: 0-8218-3988-8.

[8] Lopez R, Special Weingarten surfaces foliated by circles, Monatsh Math. 2008; 154(4): 289-302.

[9] Milin-Šipuš Z, Divjak B, Surfaces of constant curvature in the pseudo-Galilean space, Int. J. Math. Math. Sci. 2012.

[10] Öztekin HB, Tatlipinar S, On some curves in Galilean plane and 3-dimensional Galilean space, J. Dyn. Syst. Geom. Theor. 2012; 10(2): 189-196.

[11] Pressley A, Elementary Differential Geometry, second edition. London, UK. Springer-Verlag London Limited, 2010.

[12] Ro JS, Yoon DW, Tubes of Weingarten Types in a Euclidean 3-Space, Journal of the Chungcheong Mathematical Society, 2009; 22(3): 359-366. doi: 10.14403/jcms.2014.27.3.403

[13] Röschel O, Die Geometrie des Galileischen Raumes, Bericht der Mathematisch Statistischen Sektion in der Forschungs-Gesellschaft Joanneum, Bericht Nr. 256, Habilitationsschrift, Leoben, 1984.

[14] Savci UZ, Gorgulu A, Ekici C, Parallel surfaces of ruled Weingarten surfaces, New Trends in Mathematical Sciences, NTMSCI 2015; 3(4): 237-246.

[15] Šipuš ŽM, Ruled Weingarten surfaces in the Galilean space, Periodica Mathematica Hungarica 2008; 56: 213-225.

[16] Yaylı Y, Sağlam D, Kalkan Ö, Parallel Linear Weingarten surfaces in E^3 and E_1^3, Mathematical Sciences And Applications E-Notes, 2013; 1(1): 67-78.

[17] Yoon DW, Surfaces of Revolution in the three Dimensional Pseudo-Galilean Space, Glasnik Math. 2013; 48(68): 415-428.

ISSN: 2153-8301   Prespacetime Journal   www.prespacetime.com
Published by QuantumDream, Inc.

Opinion

On the 2020 Physics Nobel Prize

Ervin Goldfain[*]

Advanced Technology and Sensor Group, Welch Allyn Inc., Skaneateles Falls, NY 13153

Abstract

The purpose of this note is to draw attention to several open issues related to Penrose's singularity theorem of 1965.

Keywords: Singularity theorems, incomplete geodesics, Cauchy hypersurface, trapped surfaces, black holes, general relativity, quantum gravity.

The distinguished theorist Roger Penrose has recently shared half of the 2020 Physics Nobel Prize for his theoretical work on the Black Hole (BH) singularity problem and its connection to General Relativity (GR). The formal announcement of the Nobel Prize Committee acknowledges that *"Penrose's discovery of the singularity theorem showed that black holes are a robust consequence of the theory of general relativity, forming naturally in very overdense regions."* [1].

In our view, it is important to bring up several lesser known aspects of Penrose's singularity theorem that remain open for debate. In particular,

1) In general, the description of BH dynamics rests on the tacit hypothesis that the spacetime continuum of strong-gravitational fields conforms to the non-Euclidean geometry of *smooth manifolds*. This hypothesis is likely to fail near the event horizon of supermassive BH's. It is conceivable that the onset of ***far-from-equilibrium*** conditions and of a ***transient regime of large metric fluctuations*** falls entirely outside the low-energy framework of GR and conventional field theory.

2) The N-body problem of celestial mechanics leads to ***Poincaré resonances*** and the issue of ***non-integrability and chaos***, overlooked by the majority of models on gravitational collapse [2-4, 7]. In these dynamic regimes, the very notions of trajectory, geodesic or trapped surfaces are likely to become ill-defined, break down entirely or require significant review.

---

[*]Correspondence: Ervin Goldfain, Ph.D., Photonics CoE, Welch Allyn Inc., Skaneateles Falls, NY 13153, USA
E-mail: ervingoldfain@gmail.com

3) Penrose's work shows that inside of the event horizon the radial direction is time-like and time ends at the singularity. As a result, "space" is alleged to turn into "time" inside the singularity. However, this statement makes sense if (and only if) the union of space and time survives inside the singularity, which means that the concept of speed of light in vacuo maintains its traditional meaning below the horizon. This may not be the case if the singularity exhibits ***non-trivial topology*** and is ***no longer differentiable*** in the conventional sense.

4) Ref. [5] shows that 'The presence of a ***Cauchy hypersurface is not guaranteed*** as was shown by Penrose in another influential paper [247] —published almost simultaneously with [248]— where he discovered that plane waves in GR do not admit Cauchy hypersurfaces. Whether or not physically realistic spacetimes are globally hyperbolic is related to the so-called strong cosmic censorship conjecture". There are some indications today that the strong cosmic censorship conjecture may be wrong [6].

5) Along the same lines, [5] also remarks that "… the ***initial/boundary condition*** is absolutely essential in the theorems. Whether or not the initial or boundary condition is satisfied by actual physical systems is debatable. We will probably never know if the entire Universe —not only the observable one— is spatially finite, or strictly expanding now. Thus, the formation of closed trapped surfaces in the collapse of physical systems given some realistic initial conditions has become the main area of research to elucidate this question, see end of subsection 7.2. One can also follow a similar program, towards the past, in the case of the expanding Universe. Of course, the formation of closed trapped surfaces may depend critically on the given initial conditions, so that this is an area of active research and great relevance in numerical and mathematical relativity now, see section 6."

6) Finally, following [5], one learns that "A very important line of research arises from the ***tension between the singularity theorems and the (yet unfound) theory of quantum gravity***. It is widely accepted that the existence of classical singularities signals a breakdown of the classical theory at extreme conditions, which is precisely when gravitational quantum effects will become relevant. Thus, there is a need to clarify if the singularity theorems, or part of them, can survive when entering into a quantum regime, or if they then simply vanish altogether. For a general discussion, see [35]"

We close by quoting the summary statement of [1] which goes on to caution that:

"The extent to which the structure of a black hole surrounded by an event horizon actually match the predictions of general relativity is still an open question. Nature may still have surprises in store."

Received October 11, 2020; Accepted December 21, 2020

(Published in Prespacetime Journal| December 2020| Volume 11 | Issue 7 | pp. 637-639)

Goldfain, E., *On the 2020 Physics Nobel Prize*

References

1. https://www.nobelprize.org/uploads/2020/10/advanced-physicsprize2020.pdf
2. https://www.math.purdue.edu/~yipn/543/holmes-poincare-chaos.pdf
3. Gutzwiller, M. C. "Chaos in Classical and Quantum Mechanics", Springer-Verlag, 1990.
4. http://www.bourbaphy.fr/chenciner.pdf
5. https://arxiv.org/pdf/1410.5226.pdf
6. https://arxiv.org/pdf/1710.01722.pdf
7. https://www.ncbi.nlm.nih.gov/pmc/articles/PMC47574/

Commentary

On Complex Dynamics as Foundation of Relativistic Spacetime

Ervin Goldfain[*]

Advanced Technology and Sensor Group, Welch Allyn Inc., Skaneateles Falls, NY 13153

Abstract

This commentary clarifies the interpretation of the Gaussian random walk model of spacetime introduced in this journal [1].

Keywords: Complex dynamics, relativistic spacetime, Gaussian random walk.

As conjectured in [1], at energies near or exceeding the Fermi scale (M_{EW}), conventional space and time coordinates turn into *statistical entities* and create dynamic conditions falling outside the effective framework of Quantum Field Theory. In particular, the scaling of coordinates (X) near or above M_{EW} is determined by the following probability distribution

$$P(X,L) = X^{-\tau_X}\Phi(X/X_c), \quad X >> 1, \quad L >> 1 \tag{1}$$

$$X_c = L^{D_X}, \quad L >> 1 \tag{2}$$

The contribution of the cutoff function Φ becomes negligible when X falls far below its cutoff value or $L \to \infty$, that is,

$$\Phi(X/X_c) = const., \quad X << X_c \text{ or } L \to \infty \tag{3}$$

The key hypothesis of [1] is that the limit (3) of the power-law (2) reproduces the probability distribution of *Gaussian random walks*. Our goal here is to further elaborate on this point following the philosophy of the Renormalization Group program.

Consider a generic random walk in (2+1) spacetime dimensions having a constant step size commensurate with the magnitude of X. The probability distribution of this process replicates the power-law described by (1)-(3), namely [2]

$$P(\mathbf{X}) = \frac{1}{4\pi} X^{-2}\delta(|\mathbf{X}| - X) \tag{4}$$

where

$$\mathbf{X} = (X_0, X_1, X_2) \tag{5}$$

Next, proceed to constructing the probability distribution for the coarse-grained step

[*]Correspondence: Ervin Goldfain, Ph.D., Photonics CoE, Welch Allyn Inc., Skaneateles Falls, NY 13153, USA
E-mail: ervingoldfain@gmail.com

ISSN: 2153-8301   Prespacetime Journal   www.prespacetime.com
Published by QuantumDream, Inc.

$$\mathbf{X}' = \sum_{1}^{n} \mathbf{X}_i \tag{6}$$

in which $n \gg 1$ is the overall number of steps. The coarse-grained probability distribution in $D \geq 2$ dimensions can be derived through the cumulant expansion method [2]. The result in the 3+1 spacetime reads

$$P(\mathbf{X}') = \frac{1}{\sqrt{2\pi n \sigma_0}} \exp(-\frac{|\mathbf{X}'|^2}{2n\sigma_0^2}) \tag{7}$$

which represents a standard Gaussian probability distribution with fractal (Hausdorff) dimension $D_{RW} = 2$.

The takeaway point of this discussion is that, in the limit of exceedingly long random walks consisting of statistically independent steps, processes of the type (4) are reducible to Gaussian Random Walks. This is to say that space and time coordinates evolve according to the *same probabilistic rules* at large energies, graciously blending into the four-dimensional continuum of relativistic physics near or below M_{EW}.

There is another way of phrasing this result. Assume that space and time *are not* integrated into spacetime but rather treated as individual random variables above M_{EW}. In this case, the fractal dimensions of space and time are unequal ($D_{RW} = 1$ for time in one dimension and $D_{RW} = 2$ for two or three dimensional space). As a result, the locality principle expressed by (9) in [1] fails to hold because (10) is no longer equal to one. The unavoidable conclusion is that space and time *must* be joined into a single entity, the Minkowski spacetime of Special Relativity.

Received August 7, 2020; Accepted December 21, 2020

References

1. Goldfain, E. (2020), *Complex Dynamics as Foundation of Relativistic Spacetime*, Prespacetime Journal, 11(4): 314-317.
2. Creswick, R. J., Farach H. A., Poole C. P., Introduction to Renormalization Group Methods in Physics, John Wiley & Sons, 1992.